ROUTLEDGE LIBRARY EDTIONS:
GLOBAL TRANSPORT PLANNING

Volume 8

I0124405

TRANSPORT AND THE PUBLIC

TRANSPORT AND THE PUBLIC

J. A. DUNNAGE

Routledge
Taylor & Francis Group

LONDON AND NEW YORK

First published in 1935 by George Routledge & Sons Ltd

This edition first published in 2021
by Routledge
2 Park Square, Milton Park, Abingdon, Oxon OX14 4RN

and by Routledge
605 Third Avenue, New York, NY 10017

Routledge is an imprint of the Taylor & Francis Group, an informa business

© 1935 J. A. Dunnage

British Library Cataloguing in Publication Data
A catalogue record for this book is available from the British Library

ISBN 13: 978-0-367-69870-6 (Set)
ISBN 13: 978-0-367-74042-9 (hbk) (Volume 8)
ISBN 13: 978-0-36774-055-9 (pbk) (Volume 8)

Publisher's Note
The publisher has gone to great lengths to ensure the quality of this reprint but points out that some imperfections in the original copies may be apparent.

Disclaimer
The publisher has made every effort to trace copyright holders and would welcome correspondence from those they have been unable to trace.

TRANSPORT
AND THE
PUBLIC

By

J. A. DUNNAGE

F.I.T.A., M.J.I., A.M.Inst.T.

LONDON

GEORGE ROUTLEDGE & SONS, LTD.

BROADWAY HOUSE, 68–74 CARTER LANE, E.C.

1935

Printed in Great Britain by Butler & Tanner Ltd., Frome and London

CONTENTS

CONTENTS

CONTENTS

CONTENTS

AUTHOR'S NOTE

TRANSPORT problems vitally affect us all. The mistaken and inadequate solutions from which we suffer mean choked streets, industries overburdened by uneconomic costs, needless daily discomfort for millions of us, and perhaps the premature termination of our lives. Hence, efforts to discover and achieve the proper development and use of all our transport facilities closely concern every intelligent citizen—and he will certainly pay, in one way or another, for the mistakes.

Much has happened in transport during and since the War of which the average man has seen but small, unconnected and at times unintelligible signs. Indeed, John Citizen can only hear or read, as a basis for his personal views, such matter as is released to the press by one or other of the directly interested transport operating groups. How, then, shall he know truth from—propaganda ?

This little work seeks to outline in clear language some of the problems now occupying the minds of transport experts, and to reach a reasoned estimate of future possibilities in this sphere.

War conditions hastened transport progress and this fact, plus the intensified business competition of post-War years, hand-to-mouth buying, and housing difficulties—with many other aspects of the general sociological problem—have combined to force the pace in social reform and in radical new conceptions in business administration. They are at the same time placing a truly efficient transport system among factors vital to the national well-being.

From a mass of technical considerations there emerge certain leading thoughts of primary importance. Transport problems do not stand still. It is futile to sigh for a return to the pre-1914 position. Even if desirable—which the writer utterly denies—it is impossible. Discoveries have been made, forces harnessed, amalgamations and reorganizations planned and started which cannot be countermanded. We must go on. We are, indeed, rushing on—literally as well as figuratively.

Where are we rushing? Is there any directing policy taking care of the nation's interests, as distinct from those of competing groups of financiers? Despite months and years of costly research work by various Commissions and Committees, despite even the existence for fifteen years of a Ministry specially devoted to such studies, not enough heed has yet been paid to what the nation wants. Until the Government of the day, whatever its political colour, does hammer out an all-embracing transport policy, honestly designed to safeguard the nation's interests and increase the comfort of every citizen, changes in the law will still tend to be effected at the instance of the most clamorous and powerful contestants rather than those having the best case.

Granting that the events of the moment may loom unduly large in our eyes, and that most active people tend to feel that their particular enthusiasm is of pivotal value; yet the present time is certainly of more than ordinary importance in Britain's history.

AUTHOR'S NOTE

We are on the eve of fundamental changes as the true implications of the power age are realized and we take courage and accept the bounty which nature and science jointly have made available for us. Though our financial conjurors will fight bitterly to keep the public ignorant of the basic money swindle which to-day cramps our lives at every turn, the wool is being drawn from the eyes of more and more people, and soon the demand for a fair sharing of the nation's cultural inheritance will be irresistible. Then things will happen rapidly, and much that to-day seems visionary will become easily possible.

Detailed and elaborate forecasts may easily meet the same fate as has befallen others in the past, yet anybody whose function it is to consider plans that may take two or three generations to fructify must try to realize what conditions may be like at the end of that period. Hence transport men also may find these chapters of some interest.

Obviously, to compress this subject within the dimensions of this volume has meant dealing only very broadly with each section of the problem. Much that might have been said has been omitted. One does hope, however, that enough has been written to whet the appetite of every reader and send him to the daily press with a much more discerning eye and critical mind.

No writer attacking such a topic would wisely try to evolve a complete plan from his own inner consciousness. All knowledge is gleaned from others, and my debts to fellow-students of these problems are many. Thanks and acknowledgements are especially due to the following gentlemen, whose words have from time to time been heard or read, and whose ideas have at one point or another afforded aid or stimulated fresh thought :

On Planning and General Matters : Principal Grant Robertson and Messrs. Herbert Morrison, L.C.C., David R. Lamb, M.Inst.T., W. H. Gaunt, O.B.E., M.Inst.T., Fredk. Smith, F.I.T.A., John B. Andrew, A. C. Bossom, Thomas Adams, A. R. Polson, F.I.T.A., and F. J. Osborn.

On Railway Matters : Messrs. Kirkaldy and Evans, C. E. R. Sherrington, M.Inst.T., and H. R. Caulfield Giles, M.I.T.A.

On Roads and Road Transport Matters : Major H. A. Crawfurd, Dr. K. G. Fenelon and Messrs. Frank Pick, M.Inst.T., W. Rees Jeffreys and Hugh Miller.

On Shipping Matters : Sir Archibald Hurd, Major Frank Bustard, and Messrs. C. E. Jordan, F.I.T.A., J. R. Cowper and A. R. Kelso.

On Port Matters : Sir George Buchanan, K.C.I.E., and Messrs. Frank Brown, M.Inst.T., and T. Bernard Hare, M.Inst.T.

On Canal Matters : Mr. W. H. Curtis, M.Inst.T.

On Air Transport Matters : Major R. H. Thornton, M.C., and Mr. G. E. Woods Humphery.

JAMES A. DUNNAGE.

WESTMINSTER, S.W.1.
August, 1935.

TRANSPORT AND THE PUBLIC

CHAPTER ONE

THE NEW CONCEPTION OF LIFE

BACK in 1900—that is, some centuries back in thought —the late Lord Balfour gave an interviewer seeking a twentieth-century message this shrewd forecast : " The problem of the past century has been production ; that of its successor will be distribution." But little of the deep truth and the wide implications of those wise words could have been appreciated by most people who read them. Perhaps he who spoke did not fully grasp the import of his crisp sentence. Yet in the obvious and limited sense as well as in the broadest imaginable interpretation of the thought, Lord Balfour in 1900 voiced a profound truth.

An Age of Potential Plenty.

We live to-day in an age of potential plenty. We have conquered the elementary problems of production, though as yet we are not putting our knowledge to anything approaching an adequate or a truly fair-minded use. Concurrently with further progress along that line there is now dimly apparent to the vast bulk of civilized peoples—and sharply evident to the few discerning eyes—an outline picture of the kind of community in which we *might* be living, say, next year, and the main framework of an unselfish, neighbourly system towards which we must strive. This system clearly requires the improvement and drastic reorganization of each part of our distribution machinery to bring it in line with what the honest, common-sense citizen wants and, in this age of potential plenty, could easily have.

For two reasons the research worker in the distribution field must struggle harder and work faster than did his brother in the production field. Firstly, because his theories, recommendations and projects touch directly—or rather, can be seen to touch directly—a greater number of citizens than did the theories and experiments of those who have done so much to reform production. Secondly, because every piece of reform work in the field of production has helped force into prominence those basic problems of distribution which in more leisurely and limited times did not appear so momentous. Indeed, but for the brilliant and rapid success of the scientist in making possible the mass production of useful and desirable things the distribution problem, though truly always there, could not yet have assumed such vital importance and might conceivably have been solved more or less at leisure. Meantime the few hundred thousand people who died of starvation through the tragic imperfections of the distribution system or other thousands who, so to speak, were never born, could hardly have become very articulate or disturbing factors.

Distribution Reform Urgently Needed.

Thus the reformer in this field is far less a master of his own pace than were the earlier pioneers who

2

brought production from an affair of hand work at home to one of the modern mass-produced and standard "lines" which, being "nationally advertised", do much to destroy the individuality of our smaller towns and villages by making any one look like any other, but yet afford the average citizen a fuller and more varied life than his grandfather enjoyed.

Traffic congestion in our streets, and the hasty, ill-judged efforts being made to minimize the evil by retarding and inconveniencing everybody instead of boldly removing the causes of the trouble, adequately prove my point. Productive efficiency has developed much faster than distributive efficiency, and the main preoccupation of so many of our blunderers to-day, instead of being the facilitation of distribution, is the limiting or stopping of production.

Yet the same two forces impelling the distribution research worker to struggle hard and fast provide cause for encouragement. For those very reasons he may expect a more vigorous public support and a more ready acquiescence from industry if only he puts forward his reform plans clearly and forcefully enough to secure for them, among all the miscellaneous clamour of life, the attention they merit. This invigorating thought is not always realized by the people it should most inspire, and so one ventures to impress upon workers in this field the urgency and the feasibility of their task as a purpose ancillary to the main mission of this work, which is to outline some aspects of Britain's transport problem and suggest some solutions.

Problem Cannot be Isolated.

Any effort to settle the physical problems of distribution without constant regard to the commercial and

3

psychological aspects is foredoomed to failure. Thus, in seeking to handle the subject commonly believed to be embraced in my title, its links with all other aspects of the entire problem of distribution and, therefore, almost every aspect of our national life, have to be kept in mind. If this be thought too big a generalization one needs but point out that the raw material of one industry is the finished product of another ; while if, in retort, the objector harks back to the primary products of mining, agriculture and fishing, one replies at once that these involve the use in every case of the manufactures of others—pit props, explosives, ropes, railway wagons, farming appliances, manure, ice, crates, steam and diesel engines, and so forth—beside of course requiring transport for the useful disposal of their output.

Thus it must be understood that, right through this effort to narrate, analyse and make proposals for the more honest and useful planning and operation of the nation's transport, and the wiser organization of the physical function of distribution, I imply the concurrent study and reform of the entire half of our corporate life which consists in sharing out and consuming those goods and services which modern productive methods and plant can make available. While some of these broader implications will need to be stated as the argument proceeds, the constant effort to state all these concurrent corollaries and implications would make for a very wordy and even tautological work wherein practical value might well be sacrificed to thoroughness. The reader is therefore asked to recall these general premises and apply them to every chapter. This is a convenient course, too, for the writer, since he thus has a ready and true answer to any criticisms from the " practical man " of to-day, whose ire will doubtless be roused by some of the proposals discussed, and who may be moved to point out indignantly how impossible they are.

Wise and Prompt Use of Scientific Resources.

The guiding idea behind these suggestions can be stated very simply and without reference to any political parties. I do not picture, nor am I working towards, a " Utopia ". That is an unfortunate word which comes so easily to the pen of the cynical press-man faced with any proposal a bit beyond the everyday thoughts of the man who gets all his mental sustenance from one of the more popular dailies. My Utopia would entail a clean sweep of nearly everything in our towns and cities to-day and much of the cheap and nasty mistreatment of the countryside. Nothing so bold is here in mind. All I picture is a people alive to the marvellous powers their scientists have put in their hands, and determined to use those powers *now* to secure a decent, comfortable and varied life for everybody ; a people inspired by practical brother-hood, abhorring cant but giving short shrift to the " twisters " in any grade of society. And—most important of all—a people which has become aware of the grave fallacies of the present system by which the control of the nation's credit is left in the hands of a few private individuals, and which is in consequence in process of solving the central problem of to-day, and finding how to put nature's plentiful harvests and the liberal output of modern power production to the uses for which they were intended.

5

Given a system of life based on these simple but adequate lines, what would the British people do in the transport field?

Transport now an Essential Public Service.

First, they would surely realize that the provision of adequate transport is a *service* rather than an industry. It is, that is to say, a feature akin to the provision of pure water, light and heat, police and sanitary protection, and elementary education, rather than to the great manufacturing undertakings. (For the present purpose we may ignore the growing acceptance of the viewpoint that *all* large scale industry also must be run with an ideal of service.) Thus it would become the community's duty to undertake, or supervise the undertaking of, the provision of an adequate transport service solely upon the basis of furnishing the greatest good to the greatest number instead of upon the old-fashioned notion of charging all that the traffic can possibly bear.

Of course, this does not imply a throwing to the winds of all commercial sagacity, but it does mean that the profit factor is subordinated to the service factor. Where a clear public need exists for transport service, that service shall be provided in the most reasonable manner possible, and if the full cost cannot be met from the contributions of users of that service, it must be made good from either the general funds of the undertaking or from the funds which are the common property of the community, as the case may be. Nothing very startling, really, when the railway principle of charging what the traffic will bear, and the history of London transport for the past twenty years, are recalled.

6

To Each according to His Need!

In some quarters there is already half-hearted allegiance to this principle. It must become whole-hearted. Workmen's trains and trams are signs of the dawn of this social conception—though, when one considers the crowded, uncomfortable state of many, it is doubtful if careful costing would substantiate the complaint of some operators and politicians that they do not pay. Anyhow, the idea is not new of charging to a section of our citizens a fare commensurate with their ability to pay, and providing them with a needed service. Some railway experts will indeed maintain that the conception here advocated is, indeed, the very one that has always underlain the railway classification and rating system, though, on the truth of this, opinion among experts is at variance. But clearly, it will not be difficult to secure a wide range of lip service, at least, to this conception, and thus to gain the basis upon which the argument can proceed.

For what follows unavoidably is that transport, being a service essential to almost every citizen, and being operated to an increasing extent with that fact well in mind, must of necessity and before very long either put its own house in much better order or submit itself to still more fundamental measures of control in the public interest.

This may appeal to some as a pretty big jump to make right off; but, unless one has the mental agility to make it, how are the supposed-to-be-conflicting interests of the owners of transport services to be properly settled, either as between themselves, or *vis à vis* their users, industrial and private? But perhaps one need not waste time arguing the obvious. The

7

inexorable logic which forced all political parties to accept the inevitability of the unified control of London's passenger transport is still at work, and still making converts, of varied political colour and business interest. So the time saved through not having to urge such a conception can be used in trying to work out some details of what still remains an extremely complex problem, by no means solved even were the whole nation united in fervent adherence to the underlying conception of a transport service run by the people for the service and profit of the people.

Dangers of Political Interference.

Whether control and reorganization of transport in the national interests means, in fact, nationalization is a big question which need not yet distress us. In any case, no single solution will suit all types of transport service.

Opponents of nationalization of transport—a phrase not to be used without fuller explanation and elimination of wrong ideas, by the way—are still prone to put up their own Aunt Sally so that they may knock it down. Consider, they say, in tones of virtuous horror, the dangers of political interference ! What *are* those dangers, and are they new ones ? Do not the big transport interests already strain every nerve in political organization (known to opposing interests as " wire-pulling ") and propaganda to gain the backing of a credulous or indifferent public when they wish Parliament to do or not to do something for them ? Are not their efforts put forth chiefly for their own advantage, as distinct from that of the public ? Has anybody heard of a railway company " lobbying " or sending a brief to a Member for a railway-wagon-building constituency; or of a road transport association entertaining a group of politicians to luncheon in the hope of " putting across " their story and wishes? Apparently, so long as political action subserves the vested interests, all is well; but if there is any danger of political power being used in the interests of the general body of voters, then some such cry is invariably raised.

Anyhow, the hackneyed objections to nationalization are now rather beside the point, for the business community and the civil servants are competing in their eagerness to hammer out a new technique which may not be ideal, but is at least free from the more obvious defects that early opponents of " nationalization " pointed out. By the method of trial and error we are finding how to create and use the Public Corporation or Board, by means of which such measure as seems necessary of commercially inspired operation can be joined with direct responsibility to the nation, but without the day-to-day interference with detailed administration which might justly be feared by all concerned. We are learning, as the Rt. Hon. Oliver Stanley recently put it, " to eliminate the waste of competition while retaining its incentive ".

Varied Facilities at the Nation's Disposal.

More than this brief hint of the inevitable results of our study of transport administration problems would be premature until the problems themselves have been outlined. Let us proceed, then, to consider some of the simple natural aspects of our subject.

Since sea water is free, and canal water fairly cheap to direct and maintain where it is wanted, transport by coastal or canal craft might be expected to be quite

9

economical from a simple engineering point of view. So they are, provided departure and destination points for a given consignment are near enough to the water and the traffic is or can form part of an economic load for each of the vehicles concerned in its carriage. As things are, the various terminal costs often offset the inherent advantages of water transport. Possibly capital judiciously applied to improvements at the crucial points might remove the difficulties and make the complete route a good one of constant utility.

Engineers claim, too, that modern railway transport is very efficient, in the simple sense that up-to-date locomotives can haul a very great weight if it is placed in large vehicles carried on steel-tyred flanged wheels which run over smooth steel rails. Several hundred tons can thus be conveyed at a low cost for fuel, and under the *direct* control of only two or three men. This is true enough, though in practice the overheads go a long way to counteract those inherent advantages.

Road transport, though inherently less efficient than the methods just talked of, has the big advantage of flexibility : it can go almost anywhere and do the work the public wants at a throughout cost which often competes with the cost by water or rail. There is no true validity in the retort that canal or rail could do as well if they had the advantage of a permanent way net as complete as that which exists for road vehicles, for a public road serves many good purposes beside that of providing a highway for modern mechanical road vehicles ; it is clearly of more value to the public for these general, non-transport purposes than either a canal or a railway are ever likely to be. So that arguments tending towards the arbitrary fixation of " permanent way " costs as between canal, rail and road in the interests of a phantom called "fair competition" are singularly unconvincing except to those who insist on placing their particular sectional interest in front of the national interest.

Air transport, to counterbalance its free track, has not yet developed its technical efficiency to the greatest possible extent in weight of paying load per horse power unit. Big advances have been made, but more are necessary before air transport can be effectively compared with older systems on the simple points of cost or safety.

Yet an intelligent forecast must be attempted of the degree of progress likely in each type of transport unit during the next few years, and to obtain this both the opinions and the characters and interests of many experts must be studied and synthesized. The over-enthusiasm of the man of one idea must be set against the selfish opposition to progress of those who imagine their interests may be prejudiced by some reform. Other and grave difficulties arise when one turns from a study of what is technically the best unit for a job, to face the legacies of ancient plant or the weight of derelict capital which cannot be ignored in any practical comparison. What shall be "written off" or scrapped? What shall be "made over"? How shall the various clamorous interests be handled or disposed of? Before trying to deal with these questions or the very thorny ones arising under "division of function" we must lead the reader a little way through the past history of transport in Britain so that he may share with us something of the picture at which we are gazing when constructing our subsequent arguments.

CHAPTER TWO

BRITAIN'S RECENT TRANSPORT HISTORY

AN adequate history of transport in Britain from Celtic and Roman days would be a fundamental and illuminating contribution to the history of social, economic and administrative evolution in our country ; it would certainly be a contribution, too, to the history of science. Such a history has not yet been written.

Transport has two main functions : the conveyance of goods for exchange or consumption, and the conveyance of human beings or animals, i.e. of living organisms. Professor Grant Robertson reminds us that the flight of Israel out of Egypt was a big problem in transport, the success of which ultimately turned on a miracle ; and Pharaoh failed not so much because of the miracle as by the inadequacy of his transport engineers.

In any given society the origin and need of transport start and end with the desire or need to exchange or transfer material goods and human beings. A static and self-sustaining society has little need of transport. Norman England, producing food, clothes and habitations from internal and localized sources but with no marginal surplus, had small need for transport, except that the political accident that the Kings of England were also great rulers on the Continent meant that, with their Courts, they had to pass from Winchester to Rouen and Tours and Angers, and back again. A cosmopolitan church was another cause of transport. But a dynamic society, producing more than it can consume and driven by a dynamic mind which conceives that by exchange of goods and ideas it can alter and heighten the efficiency of life, has a need of transport proportioned precisely to the strength of the dynamic force operating on the community.

Periods of activity in scientific discovery either are periods or lead to periods of great activity in the technique and mechanism of transport. Until about 1912, transport was limited to land and water. The brothers Wright added air—for the balloon never seriously came into the field of transport science. But why (asks Professor Grant Robertson) did human society remain stationary in the mechanism of transport from Tutankhamen or Adam down to James Watt? The age of the Pharaohs had just the same mechanical means for conveying goods or humans as the age of George II. For thousands of years human legs, horses, animal-drawn vehicles, roads for such, oars and sails constituted the sole material and the limits of transport. Why did the steam engine and the railway come when they did? Why were they so long postponed? Why has the period from 1890 onward been so revolutionary and so crowded with inventions—the bicycle, the inflated rubber tyre, the electric tram, the electric tube, the internal combustion engine and the motor-car, the aeroplane, and now the substitution of oil for coal and steam? Will the rate of discovery and invention be maintained, and in what directions and with what results?

Roads in the Middle Ages.

There is much evidence of the badness or non-existence of roads from the days of Queen Elizabeth onwards. This evidence is almost incredible to-day. But from the middle of the seventeenth century we can trace slow but accumulating improvement. New roads began to be made, old so-called roads or tracks were patched and repaired. From 1660 onwards both internal and external trade steadily increased in volume. England after the great Civil War ceased to be a self-sufficing country, and began to develop a marginal surplus available for exchange. It is difficult to disentangle cause and effect. Economic activity demanded new or better roads, new or improved roads stimulated economic activity. The route became important. The surveyor came into greater prominence, either to create a new route or re-make an existing one.

As Britain is an island the route to and from certain ports was vital, and in making new routes or improving old ones the basis of a new science of road-making was laid. Bridges and the science of bridge-building came into the picture, and the way was prepared for Rennie, Paterson, Telford and Macadam. Yet improvement was limited by the available factors that determined rate of speed and comparative rate of cost. The horse, and the horse-drawn vehicle, were the limiting conditions. The breed of horses was improved; new vehicles were invented; new wagons devised; but the law of diminishing returns began to operate, and even on the best road the best horses with the best vehicles could not go beyond what horse and man could do. And the best was very expensive.

14

For goods in bulk the limiting point was soon reached. By the middle of the eighteenth century development was again halting. Something new was needed. This point was reached just when the industrial revolution was beginning to revolutionize our whole economic life, when coal, the blast furnace, and iron opened up a new era. The revolution in economic life required a revolution in transport, especially transport of goods in bulk. The application of water power to machinery created the age of canals. Watt's steam engine—steam applied to machinery—set engineers thinking, and the result was the railway, and the railway was the pioneer of a social and economic revolution.

Railways Pioneered a Social Revolution.

Even a cursory study of this period shows up a vital truth, which is at the heart of the impulse toward the writing of this book. All transport developments create crucial problems of public administration. Who is to maintain and who is to pay for improved transport ? In seventeenth-century England the great obstacle to improved roads was the refusal, say, of a parish in Surrey to pay for or maintain a bit of a road or a bridge which would benefit agriculturists or manufacturers a hundred miles away. England's transport history for a century and a half after 1660 is filled with experiments in trying to secure adequate maintenance and to distribute costs. We get compulsion, monopolies, turnpikes (i.e. makers of a road taxing users), statutory rates, free competition, anarchy, bankruptcy—in fact, we run through the gamut of human versatility in the devising of administrative machinery and fiscal expedients, and in the art of

15

placing a burden on the back of somebody else. To-day we are confronted with the same problem. But we do see in this bewildering jungle the slow and steady clearing of a path, and the slow emergence of new principles of taxation and finance.

It is sad to reflect, however, that from earliest recorded times military requirements influenced design, both of land vehicles and ships, and led to their improvement. In pre-Christian times the finest example of the shipbuilder's art was the fighting galley, while the war chariot was the fastest land vehicle in the days of Julius Cæsar. Roman roads in Britain were built to provide speedy communication with garrisons and outposts. It is commonplace that both aircraft and motor vehicle design made more progress between 1914 and 1918 than they would have made in four times as long but for the Great War. Nor would the change from wooden to iron and, later, to steel ships, have been so rapid if warship designers had not realized that metal was a better protection against enemy gunfire than the wooden walls of the opposing fleets at Trafalgar.

When Government fostered Coasting Trade.

The Stockton and Darlington was the first public railway in the world, and coal transport was its primary purpose. From 1800 to 1825 the only means of inland goods transport except the roads, which were unsuitable for heavy merchandise, were the canals. Over 2,000 miles of canal were in existence before the first public railway, and they carried much passenger traffic ; though merchandise, especially coal, had also been responsible for their creation. Incidentally, in 1800 only one canal—the Grand Junction—entered London,

16

for the Government forbade coal to be carried to the capital save by sea. This was to encourage the Newcastle coastal trade, then regarded as a "nursery of seamen". So there is a precedent for the Minister of Transport should he feel it wise to lay down any general instructions to transport people for a "division of function".

The public opening on 27th September 1825 of the Stockton and Darlington Railway opened the public's eyes to new possibilities, since for the first time some 600 people in 22 trucks and 12 truck loads of coal and flour were safely hauled by one locomotive at an average speed of 12 miles per hour. Railway construction was actively pursued for twenty years, and, after the settlement in 1846 of the "Battle of the Gauges", a period of consolidation followed, and would have continued further than in fact it did but for Parliament's decision that while an amalgamation of continuous line might be safely allowed, one of competing lines would be against the community's interests.

The railway companies' early idea of acting, like most canal owners, merely as toll takers, permitting vehicle owners to operate over the rails after paying a toll, was soon abandoned, for it became clear that the pace of vehicles running on rails was set by the slowest. To meet the difficulty sidings were installed, but it was apparent that to get good results a railway must own, control and maintain the permanent way, and also fully control all the services functioning along the line. Not only must the railway company own a road, but it must be a common carrier and have responsibility for both goods and passenger trains running on its system. This greatly changed the ideas of how a transport service should be conducted, and soon made it clear that the new kind of transport system could handle large quantities of traffic, if regular and for a through haul, with but little handling and at a low cost.

Using the Political Machine in 1825.

It is recorded that the Stockton and Darlington Railway counted on making its main profit from carrying coal, chiefly for shipment from Stockton to London, and that Mr. Lambton, M.P. for Sunderland, secured an amendment to its bill while in the House of Commons so that for hauling coal to Stockton the maximum rate should be $\frac{1}{2}d.$ per ton per mile. Though the Company were granted a maximum rate of $4d.$ per ton per mile on coal carried along the line for local sale, Mr. Lambton's amendment was adopted. He had framed it in the interests of his constituents, whose coal largely supplied London, confidently expecting that the low maximum rate would prevent any coal passing along the line for shipment at Stockton. To the general surprise, however, this traffic proved the main dividend-earning business of the railway company, and soon more than 500,000 tons of shipment coal were annually being carried.

Early railway promoters had to surmount or buy off much prejudice and selfish opposition, and railway capital still suffers through the enormous sums so paid away. The stories of railway speculation and financial disasters are now well known, but from 1845 some order began to emerge, and Parliament, after being so congested with railway business that a stoppage of the law-giving machinery seemed likely, adopted a definite policy as to concessions. The

18

Railway Clauses Consolidation Act, 1845, required all railway bills to be presented in a set form and limited the attention which the House needed to give to each. The amalgamation process gradually continued, but bitter enmity also existed between certain companies, and attempts were made to "freeze out" smaller and less-favoured companies. Since free competition was held as infallible, Parliament viewed with concern the evolution of large companies and feared the results of this "freezing-out" policy. Yet nothing definite was done until a Committee, on which sat Gladstone and Cardwell, in 1853 invented "compulsory running powers" which were embodied in the Railway & Canal Traffic Act, 1854. Under this Act, specific clauses could be inserted in railway companies' Acts giving them power to run over other people's lines where, otherwise, a connecting link would be missing. By the same Act, all railway and canal companies were required to "afford all reasonable facilities" for dealing with traffic from all other companies. So, after Parliamentary approval, any company could get access for its trains to the system of any other company, and the "freezing" process stopped—a fact of economic importance, for it contributed to the extension of transport facilities.

Companies negotiated "friendly" arrangements for running powers though, as Sherrington says, not always for purely friendly reasons. Such powers at times were granted to rivals for fear that otherwise they would embark upon new and vigorous competition ; it being deemed better to let a rival into a town and have some control of his traffic through handling it, rather than leave that rival to make his own way in. When the old Midland Railway used the Great

Northern line between Hitchin and King's Cross the traffic outgrew the facilities. The Great Northern gave their own traffic preference, and Midland trains were so constantly delayed that in the end the Midland Railway in 1863 had to secure an Act and build their own London terminus.

Regulation of Railways in Britain.

To deal with disputes under the 1854 Act clauses which required the railways to afford " reasonable facilities " some Tribunal was needed. The Act's framers wanted to entrust this power to the Board of Trade, but it seems the railway interests were strong enough to prevent this, and the Court named in the Act was the Court of Common Pleas, whose judges declined to deal with purely technical questions. Thus for nineteen years an important purpose of the 1854 Act was stultified, and those who saw the weakness failed to gain their purpose. At length, in 1873, Railway Commissioners were appointed, and to them were transferred the powers given by the 1854 Act to the Court of Common Pleas. Of value to traders also was the empowerment of the Commissioners to hear and determine questions as to terminal charges (where not fixed by Act of Parliament) and to make orders requiring the companies to divide up particular charges so as to show the toll, the conveyance charge and the terminal charge. The powers of the Commissioners were further increased, but a most important advance was the Railway and Canal Traffic Act, 1888, which fundamentally affected rates and charges and created a new and permanent body called the Railway and Canal Commission, composed of two appointed and three ex-officio members (judges of the superior courts of England, Scotland and Ireland). The other two members were appointed by recommendation of the Board of Trade, and "one of them shall be of experience in railway business".

The Railway and Canal Commission.

By the 1888 Act the Commissioners could order the companies to provide traffic facilities despite any agreements into which they might have entered. They also were given supreme authority to deal with questions of rates and facilities, and could be appealed to by a trader with a grievance against a particular railway for charging excessively for services rendered, or for unduly preferring a competitor in the matter of rates. But the Commission did not satisfy the trading community, for it was too costly. Since railways always briefed eminent counsel, traders felt bound to do the same, and in practice only large concerns could try to use the Commission machinery to rectify alleged grievances.

As big companies controlling trunk lines emerged, they long remained in vigorous competition between many points ; and although all worked to one merchandise classification and to maxima schedules of rates imposed by Parliament, writers of the period suggest that there was much room for competition in the matter of facilities. Passenger services were duplicated, and the companies tried to beat each other in speed and comfort of travel. In the goods department it is suggested that indirect preference could be and was often given by generosity in meeting claims for damage, misdelivery and delay, or laxness in enforcing charges for warehousing goods or for truck demurrage. In time the railways tired of this

21

policy and began to act together to enforce their supplementary charges and resist claims. A vigorous outcry from traders was not very successful, and the companies continued to draw closer together until at length several working agreements were arranged, such as that between the London, Chatham & Dover and the South Eastern, and one between the L.N.W., Midland, and Lancs. & Yorks. Railways. In the first example, receipts were pooled and net profits divided on a proportional basis. In the second case, only receipts from competitive traffic were pooled and divided proportionally, while to an extent the actual handling of the traffic was pooled—e.g. joint vans for collecting and delivering parcels.

There was also a very close working agreement between the Great Central, Great Eastern, and Great Northern companies, and the last two had a working arrangement with the Great Western at certain points. The Great Western and London & South-Western had a working agreement, and so had the Scottish companies.

Eliminating Inter-Railway Competition.

This movement towards eliminating competition being the voluntary work of the railways was, presumably, to their benefit, for it reduced working expenditure by abolishing duplicate services. Some passenger trains were cancelled and the cost of canvassing for traffic was said to have been reduced. The companies themselves said that such economies had become desperately necessary. As Kirkaldy and Evans point out (*History and Economics of Transport*, Pitman, 5th edition, 1934) the circumstances of their development in the last two decades of the nineteenth century had added enormously to their capital accounts, and frequently the expenditure was upon enlargements which, while imperatively necessary for the handling of traffic, did not result in an increase of earning power or a decrease in working expenditure proportionate to the capital expenditure. Further, there had been a steady increase in operating expenses, and these things were impairing their ability to maintain dividends. To note that these points were made by the learned authors in the first edition of their standard work in 1915 is to discount heavily the clamour made by the railways in 1928 and onwards, when their troubles were repeatedly attributed to road transport competition.

However, the Great War of 1914–18 radically changed many outlooks. Whereas hitherto there had been a voluntary drift by the railways toward closer working agreements, while Parliament more or less consciously sought to maintain transport competition and guard the public against the abuses of monopoly, war conditions led to a big effort to wipe out, for the time, in the national interest, all rivalries and minor differences of management, control and traffic manipulation on the railways, and to the operation of the service as one coherent whole. The possible need for this step had long been foreseen, for by Section 16 of the Regulation of the Forces Act, 1871, the Queen might by Order in Council declare that an emergency had arisen in which it was expedient for the Government to control all railroads, and the Secretary of State was then empowered by warrant to authorize any person or persons to take possession of the railways and use them for Her Majesty's service at such times and in such manner as the Secretary of State might direct.

Railway Unification During the Great War.

So in August 1914, by virtue of these old powers, the railways were put under the control of a Railway Executive Committee, established in 1912 for the purpose, and consisting of the General Managers of the chief railways. The Government agreed with the railways that, as compensation for all the special services performed for military and naval transport, they should be paid the sum by which the aggregate net receipts of the railways while in Government possession fell short of the aggregate net receipts for the corresponding period of 1913. This basis was modified to allow for a proportion of war bonuses paid to railway employees being met by the Government, and also, as the war was protracted, to provide for necessary works of renewal and extension before arriving at the balance of net receipts, even though such renewals and extensions had not, in fact, been carried out. Yet the main principle unimpaired was the setting up of a standard of net receipts and the promise of the Government to contribute the amount by which the actual profits fell short of this standard ; and this despite severe restrictions on normal civilian travel.

War control broke down the remaining barriers between the companies, for the Railway Executive Committee had unquestioned authority to carry out the transport needs of the War Office, getting traffic over the quickest or most convenient route quite irrespective of its original ownership. This experiment in the operating possibilities of unification must have led to a big change of mind for many senior railway officers, so paving the way for the reorganization of railways which was the first big proposal of the newly created Ministry of Transport soon after the finish of the War.

The Ministry of Transport Created.

This new Ministry, the work chiefly of Sir Eric Geddes (Earl Haig's Director General of Transportation in France in 1917–18), took from the Board of Trade all its powers affecting railways and canals, and also was given general supervision of roads and road transport. Still wider powers were visualized in Geddes' original Bill, but vigorous opposition led to the withdrawal of clauses concerning the control of ports and coastwise shipping.

The vast scheme of railway amalgamation took shape in the Railways Act, 1921, by which 121 companies were amalgamated into the four now familiar groups. Two important principles laid down by the Minister of Transport were not adopted when the Bill took shape as an Act, and are now conveniently forgotten by most people. It may be well to drag them out.

In an intensely interesting White Paper the Minister foreshadowed the 1921 Act and outlined its policy for the regulation of railways by the State. After explaining the grouping proposals he turned to *Management*, and emphasized the Government's view that the time had come when the workers—both officials and manual workers—should have a voice in management. Of the Board of not more than 21, representatives of shareholders should form a majority, and a proportion should hold large trading interests. The balance should be employees—one-third leading administrative officials of the group, to be co-opted by the rest of the Board, and two-thirds elected from and by the railway workers.

Equally important were the financial plans. Standard rates and a standard revenue were foreshadowed. With due care and economy the companies should be able to improve on their pre-war return. In that event the surplus should not accrue entirely to the companies. The State would be very materially extending the " Charter " of the companies *and was entitled to participate in such surplus revenues*. The Government's share would not be thrown into the general revenue of the country, but would form a Development Fund for assisting backward districts, developing light railways and other appropriate purposes connected with transport.

The work of the Minister of Transport on this point had been preceded by a certain amount of work by an almost forgotten Royal Commission set up before the War to study once more the relationship of our railways to the State. In August 1914 its uncompleted work was adjourned, but just four years later a Select Committee was set up to consider what steps ought to be taken to improve the internal transport facilities of the United Kingdom, especially by the co-ordination of the various means of transport. Two reports from this Committee were published at the end of 1918, and in 1919 Parliament passed the Ministry of Transport Act, which owed a good deal to those reports.

Growth of Road Transport.

The 1913 production of motor vehicles in the United Kingdom was but 34,000. War conditions helped to double this annual output, and after the Armistice many War Department surplus vehicles were sold

26

quite cheaply, for civil work. The 175,000 miles of road (in England, Scotland and Wales) which before 1914 had been ample for the transit of bicycles and horse-drawn carts and carriages soon became great arteries of travel. Sherrington points out (in *A Hundred Years of Inland Transport*, Duckworth) what a mystery it will be to future generations that the framers of the Railways Act of 1921 appeared to take no account at all of the growing importance of the motor vehicle and the internal combustion engine; the more so since the first Minister of Transport himself, as Director-General of Transportation to the British Armies in France, had seen the remarkable ability of the motor vehicle to handle troop movements in the field, and feed whole divisions day after day without failure.

In January 1920 and again in September the Government materially raised the level of railway charges, and though the taxpayers' load may thus have been lightened, much passenger and freight traffic was certainly diverted to the quickly extending services offered by road vehicles. The railway managements, contrary to popular belief, were not altogether blind to the motor vehicle's possibilities, and in 1922 bills were introduced by the four railway companies seeking powers to operate without restriction on the roads. After long hearings before a Joint Select Committee the railway companies very pettishly and foolishly withdrew their bills, ostensibly because the Minister of Transport sought to include clauses providing that their rates for carriage by road should not differ from their rates between the comparable points by rail. What a short-sighted action that withdrawal was for the railways—and how providential, many people feel, for the public—has since been painfully brought home to the railway leaders, who have certainly suffered for the unbalanced attitude of their advisers of the time. Surely a little judicious bracketing of those projected powers with their existent powers to quote exceptional rates was all that was needed for the railways to have accepted a clause in their bills to meet the Minister's point while yet meaning, in practice, just nothing?

Railway Acumen at Fault.

Anyhow, the railways withdrew their bills, and for nearly five years did little to improve their competitive situation. True, the general level of railway rates was forced down a bit by ordinary economic pressure, and there were stirrings among the more energetic railway minds so that containers—long ridiculed by most railwaymen—suddenly captured their interest. But during those important years for motor transport the higher railway command seemed to pay most attention to the scrupulous calculation of a theoretical " standard revenue " (never, in fact, to be attained) or the skilful and determined defence of rate relationships such as that deemed to exist between goods in various classes moving from stations in the Birmingham and South Staffordship group of stations to the principal competing ports on the Humber, Mersey, Bristol Channel, English Channel and Thames.

Though railway experts were fascinated with the minutiæ of these and kindred problems, the close observer can detect no really workmanlike handling of the larger policy questions until about 1927–8 when, at length seriously alarmed at road motor transport's progress, the railways again sought to get road powers for themselves. Once more there was a long hearing by a Joint Select Committee of the two Houses, and a great many objections, much evidence and skilled cross-examination, all costing money. Hauliers were not then organized and so could not be effectively articulate, but the leading organizations of traders were almost unanimously in opposition, for they feared a return to what they termed the old evil days of monopoly of internal transport if the railways gained these powers, concentrated new fleets of lorries on a given area until all " independents " were driven out of business, and then passed on in turn to other areas.

Clever clauses which permitted the taking of a financial interest in existing road transport concerns probably proved a winning argument ; and against the opposition of industry, continued to the last in what was regarded as a record private bill hearing (34 days), the railways were granted the powers they sought, under the form of the Railways (Road Powers) Acts, 1928.

Railways gain Road Powers.

With the long period of apparent inaction, followed, after quiet confidential work, by the dramatic announcement of the purchase of material financial interests in selected passenger transport concerns in each area of the country, one cannot here deal adequately. These interests now cover over 15,000 passenger vehicles, and the secret history of that period would fill several books and might be fascinating but perhaps libellous. But how to use their new powers in the goods sphere was a more complex problem for the railways. Overtures were certainly made to them by many struggling haulage concerns which saw in the new situation a personal ray of hope. But these were not, for the most part, the class of concerns that appealed to the railways. Firms into which they might have liked to push their way were mostly not then ready. So a simple wholesale peaceful penetration was not then possible. Rather were the railways driven slowly to act on lines which, so far as they go, every fair-minded person must endorse. They tried to rearrange their facilities so that each was used to better advantage. While seeking to keep on the rails as much long-haul work as possible, they steadily applied plans by which lorries did the auxiliary short-haul work over a somewhat wider range than hitherto.

This may sound trite, but it did mean new thought in railway practice, for it was nothing less than the real beginning of a serious attempt to give the public an attractive service that it wanted instead of merely continuing the earlier " take it or leave it " attitude responsible for so much bad feeling. It has meant the great development of railhead warehouses, to which stocks are moved from the factory in full truck or container loads, at favourable rates, and whence small deliveries are made by lorry over a radius which varies up to about 30 miles. Country lorry services now cover vast stretches of the countryside, the services being based upon the bigger towns where there is pretty good storage accommodation. In some cities the parcel delivery area has been much extended, and in others the railways now undertake point to point cartage within the urban area (i.e. handling traffic which does not pass over the railway at all). The railways own and directly operate nearly 7,000 motor lorries, and until recently they have employed on the roads 28,000 horse-drawn vehicles on which no tax is paid. These latter are, however, steadily being replaced by motor vehicles, especially of the " mechanical horse " type. Several important road goods transport concerns are now partly owned by the railways, and when the railway officers cannot secure the traffic for handling over their metal tracks it is increasingly the policy to try to secure it for their own or their allied road transport. The economics of such internal competition are open to question, except that it may tend to keep in railway hands the goodwill of a particular client.

The railways, however, still carry a load of problems from which they are with difficulty extricating themselves. Mostly these problems are mental, but not entirely so. The overhaul of railway routine called for by the Railways Act, 1921, embraced a thorough revision of the goods classification, the standard conditions of carriage, the schedules of standard rates, and provided for the computation of " standard revenues " based on actual working results for 1913 (a peak year) plus allowances for all new capital introduced since that date. The idea was that the total sum so computed was the revenue to which the railways were " entitled ", and, if they got more, part of the surplus should be returned to traders by way of reduced rates.

Those who drafted these clauses and calculated the millions of pounds forgot three things : (1) the steady downward tendency of prices during the significant years, (2) that any serious dislocation of trade would upset the validity of their reckoning, (3) how the rise

31

of a new transport element might spoil all their theories. In truth these three factors have so interacted that the theoretic standard revenue has never been attained. Each year from 1928 (when the scheme began) the actual net revenue has been much below the standard.

The Royal Commission on Transport.

For propaganda reasons, the railways have been wont to blame road transport for all their troubles. Such an attitude is not warranted, but of course there is something in it, and we must now review a few of the recent happenings in connection with the national economic problems affecting mechanical road transport. Crediting the reader with a good man-in-the-street knowledge of the development of the internal combusion engine and some of its results, we will bring him right along to 1929, when an influential Royal Commission on Transport, presided over by Sir Arthur Griffith Boscawen, began sittings which continued at intervals for over two years. Experts of many kinds, delegations from all sorts of interested bodies and private enthusiasts on particular problems appeared before this Commission, usually submitting written statements upon which they were permitted to elaborate and to face cross-examination by each member of the Royal Commission. Two lengthy reports emerged and most of the findings were translated into law (in the Road Traffic Act, 1930) before the Commission had finished its hearings, while the Third and Final Report of the Commission appeared in 1931. This dealt with the most difficult question of co-ordination between competing systems of transport, and examined very carefully the matter of the cost of the roads and the yields of vehicle and petrol taxes.

In rough figures, the Royal Commission found that the country's roads cost sixty million pounds yearly to maintain and develop. About two-thirds of this was found by taxpayers as a whole and one-third came from the Road Fund (i.e. from the payments of motor transport owners). But road transport, it was found, contributed actually a very much larger sum, which was retained by the Chancellor and applied for general revenue purposes. Altogether the road vehicle owner was found to pay by way of fuel tax and licence duty combined between fifty and sixty million pounds a year.

The Royal Commission's view was that in future one-third of the cost of the roads should fall on the taxpayers and two-thirds directly upon all motor users, so that in 1931 the figures of twenty and forty million pounds respectively were authoritatively recommended as the proper contributions of the two classes of payers.

During 1931 these recommendations were being digested and argued over by all relevant bodies. No step was taken to implement them. Apparently the views of this strong, impartial and distinguished tribunal, which had heard all sorts of experts and cross-examined them, beside checking up their submissions against all available statistical evidence, did not please the railway companies, for they did not show much prospect of any such penal imposition on independent road transport as would be likely to drive it out of existence. So that, directly the political crises of 1931 had abated, a new and clear-cut propaganda was started by the British railways having a definite and frankly stated objective, as witness the following from the railway statement to the Minister of Transport in January, 1932 :

> If experience should prove that the diversion of traffic from the railways to the road still continues, even with abated vigour, it will be necessary to consider other possibilities and to impose further burdens on the road industry which will have the effect of definitely adjusting the balance in favour of the railway industry.

Railways Dislike Royal Commission's Findings.

Some two months after the issue of that statement a weak, harassed Minister of Transport sought to avoid an awkward pressure from the railways by setting up yet another Committee, ignoring in so doing the Royal Commission's findings of 1931, still awaiting action in his office. The personnel of the new (Salter) Committee came in for grave criticism in traffic circles. This is no slight on the individual characters of the nine men appointed to it, or the honesty of their efforts to cover their terms of reference in the limited time allotted them. But a great weight of opinion at the time held that the whole idea of the Salter enquiry was misconceived, and the suitability of the Commission members to carry through the task and so the validity of their finds was sharply challenged by many. The Committee, which consisted of the four railway general managers on one side and four representatives of goods road transport on the other, with an independent Chairman (Sir Arthur Salter, K.C.B.), was on 11th April 1932 instructed

> to consider the facts relating to the incidence of highway costs in relation to the contributions of the different classes of mechanically-propelled vehicles ; to consider

34

the nature and extent of the regulation which . . .
should be applied to goods transport by road and by
rail ; and, in the light of any conclusions reached under
these heads, to make further recommendations . . .
designed to assist the two sides of the industry to carry
out their functions under equitable conditions which
adequately safeguard the interests of trade and industry ;
and to report by the end of July (1932).

One notes that trade and industry was not represented, as such, though one " road " representative
was, in fact, the Distribution Manager of a well-known
catering firm and another was a Scottish industrialist
who represented the Scottish Commercial Motor
Users' Association. Obviously, if opinions differed,
the votes of these two gentlemen would automatically
be opposed by four or even six votes ; further, they
were not able to call on the incessant services of such
a varied and highly paid " general staff " as was
behind the railway general managers throughout the
sittings of the Committee. To appoint a joint committee of two actively fighting contestants to decide
how *one of them* should let himself be mutilated would
be as sensible ; yet, to the general amazement, the
report which emerged was signed by all the Committee ! A veritable furore greeted the appearance
of this report in August 1932, and certain signatories
felt impelled to issue statements in their own defence
—an unusual thing, surely, in these matters ? Spokesmen of seventy national trading associations met,
in October 1932, at the Royal Automobile Club, and
co-ordinated their views which appeared in book
form as *The Response of Trade and Industry to the Salter
Report*, a most outspoken and scathing document which
disagreed with practically every one of the " Salter "
recommendations, and still deserves careful study.

Surprising Recommendations of Salter Committee.

Space prevents elaboration either of the detailed argument of the Salter Report or the replies of the critics. Briefly, the Salter Committee recognized that part of the annual cost of sixty million pounds for road maintenance and renewal should fall on the general community (i.e. by rates) because of the factor of " community use " of roads, which they agreed was material but on which they could put no financial value. They then created a new factor which they called " legacy from the past "—the ability of motor vehicles to-day to use roads which in part were paid for by previous generations—and on this again they did not pretend to put any specific financial value. Next they deliberately set aside, as outside their scope, the extent to which the community benefits which come to us all as a result of the accumulated experience and efforts of our forefathers should be regarded as a kind of common legacy ; and they decided to ignore the policy reasons why new systems of transport should not be discouraged by heavy taxation, even should such for other reasons seem fair. These, they said, were matters for Parliament and not for them.

So, having just mentioned, without trying to evaluate, the " community use " and having brought in also and perverted the Douglas conception of " legacy from the past " the Committee coolly proposed that these two quite unquantified conceptions should be said to cancel each other out, so leaving quite fortuitously the figure of sixty million pounds as the share of road costs to be borne by all road vehicles except horsed vehicles, bicycles, etc. !

36

Starting with these gigantic assumptions the Committee went into several places of decimals in their later desire for accuracy. After examining and rejecting various plans they advised a basis of allocation for this sixty million pounds among different classes of road vehicles, computed equally upon ton-mileage and on petrol consumption, with certain corrections to allow for speed and unlimited franchise at one end of the scale and for the extra wear and tear alleged to be caused by heavy vehicles at the other end. On the chosen basis £23,500,000 was allocated to commercial goods vehicles and £36,500,000 to all other mechanically propelled vehicles.

Regulation of goods motor vehicles by the grant of operating licences to all owners, whether hauliers or ancillary users,[1] was urged : such licences to be conditional on the payment of reasonable wages and the observance of proper conditions of service for the driver, and on an undertaking to keep the vehicles in a proper condition of fitness. The licensing authority should grant licences requested by a haulier unless it would be against the public interest because of (a) any excess of existing transport facilities ; (b) any actual or prospective congestion or overloading of roads. It would act under the Minister of Transport, who would be advised by a new permanent Central Advisory Council to be set up. Ancillary users would be granted such licences as they sought, but would be subject to conditions (a) and (b) just stated. After due notice, they would be prohibited from carrying

---

[1] This term has recently been used to cover the manufacturer, distributor, shopkeeper, etc., who owns and operates one or more motor vehicles for the movement of his own goods or goods he has sold.

37

any goods other than their own—whether for reward or reciprocal service.

In allocating the £23,500,000 (£2,500,000 more than goods road transport was then paying) among kinds and weights of vehicle the Committee produced voluminous statements showing the possible yields at ·383092517d. per ton mile and 12·617835144d. per gallon ! Yet despite the terms of reference which instructed the Committee to consider the nature and extent of the regulation to be applied to goods transport by road *and by rail* they had nothing at all to say about improving the railway services or facilities.

What a clear case of the well-trained spokesmen of a vast vested interest telling their doubtless equally well-trained politicians just what they wanted done and quietly keeping them up to the task till they had done it ! Next year (1933) a further Road Traffic Bill was prepared, to implement the goods vehicle regulation part of the " Salter " advice ; and in the same year most of the proposed extra taxation was put through the political machine.

Flat Rates Charged—and then Legalized.

Meantime an awkward situation had arisen when the railways, quoting a " flat rate " for certain traffic of a big industrial group to secure it all to rail, found the Courts going against them on a challenge to the legality of their departure from normal railway-rate principles from traders who claimed to have suffered in consequence. " Nothing easier " openly said the railway folk. " If the law has gone against us, we must change the law." And so a somewhat incongruous part was tacked on to the Bill then being drafted, and it became the " Road and Rail Traffic

Bill, 1933," which, after courageous but perhaps ill-disciplined because badly staffed efforts at mitigation by some trading and some road haulage interests, became law towards the end of 1933. It is by virtue of these powers that the railways are now prepared to quote " agreed charges " which are virtually contracts to handle all or a stated part of a business concern's traffic for a given period at a " flat rate " to any point within the stated area, which in some cases is the whole country.

Under this Act, too, the Traffic Commissioners set up in 1930 to deal with passenger vehicle licensing were also given supervision of the new goods vehicle licences, and railways were in fact given an ample chance to pry into the business of individual hauliers and use their vast resources against them in the Traffic Commissioners' Courts, to the great detriment of some hauliers.

CHAPTER THREE

THE CURRENT TRANSPORT CONTROVERSY

In the transport sphere, as in almost every other, the British nation is in the throes of a most difficult series of problems and will be forced soon to make further important decisions. Unless we can consciously make and enforce these decisions on the directly interested parties, they will shape matters to suit themselves. That may, or may not, suit the nation; we must try to determine. What are these problems? View them before the background sketched in Chapter Two. Mostly they are wrapped up in the simple yet vital question—competition or co-ordination? Everybody concerned to-day in any section of the transport profession or in industry *must* be facing some specific aspect of that central problem.

Hitherto the uninformed passenger or trader has looked on the existence and continuance of competition in the provision of transport service as of direct benefit to him; and but a slight excursion into history gives at least a surface justification for the belief. Railways brought canal rates down a century ago; pirate buses forced lower fares and improved vehicles upon the bigger bus companies in some places; canals, coastwise shipping and road transport have of recent years and in various directions provided such effective com-

petition as to modify in part the railway rates structure ; private wharves are a curb on public port
charging desires ; ample " tramp " tonnage keeps
freights low and somewhat limits the freedom of the
conference steamship lines to charge as they choose ;
railways could for a time be set in competition with
each other for certain traffics and in some areas, to the
immediate benefit of the pockets of consignors ; the
ancillary users' road vehicle, or his power to acquire
one, limits both road haulage charges and railway
rates. To the extent to which rapacity is the only
way to make profits, competition is the best curb
upon it.

On the face of things, then, the public appear justified in pressing for the continuance of such a state of
affairs as will give them the benefits of competition ;
and many responsible people—both politicians and
business folk—seem grimly determined to shout on
every possible occasion in defence of this particular type
of freedom, which they imagine they to-day enjoy.
Are they wise, and, in any case, has their desire a ghost
of a chance of true fulfilment ? Or is the conception
of competition in providing transport facilities an
obsolescent one, which people bearing political labels
of all kinds will shortly abandon ? Will the views
and wishes of the ordinary man, anyhow, have much
to do with it ? These are the real issues now needing
thought, and bound up with them are the equally vital
questions as to the type and nature of control—private,
autonomous, municipal or national—most suitable for
our transport services, assuming further basic changes
are to come. Linked also with these come questions
of the relations between the labour and supervisory
grades in transport service and the directional or own-

ing body—the many vexed points of wages, conditions of service, share in control, etc., which have hitherto been varied quite empirically as circumstances dictated, but which deserve real study in concert with the other basic questions. On the other side, the transport services must look outward, toward the bodies that in fact employ them, and, however they be managed, will need to review their " public relations " side so that their liaison with the public and industry is such that they are interpreting—more, are even hearing and considering—the views of their masters, the public, about how they should be served.

It is my case, and I come to the point without equivocation, that whatever benefits competition has brought to some people in the past, it is quickly becoming out of date in transport. Though the reader may be reminded here of the broad premises upon which this book is founded, their protection may easily be discarded, for we find to-day persons of widely differing political faiths, including many who do not seem to wish for radical changes for the better in life, giving endorsement to my statement by acting upon it. In London passenger transport the hand of the co-ordinator can be clearly traced for the past twenty years, and every step of the financial pooling and later consolidation shows that, if only for reasons of business shrewdness and despite all earlier instincts, the responsible leaders are working to the conclusions I have stated.

Questionable Benefits of Competition.

When the average man discusses competition or co-ordination of transport much muddle-headedness is likely. He speaks chiefly from surface impressions

42

of his apparent personal interest ; seldom from a deep study of public policy. He means that he would like to find a bus at the corner of his street whenever he goes there, and cares little for the economics of the problem of providing him, and millions of others, with a service that meets and even anticipates his real needs while paying as much regard to efficiency of " load factor " as conditions permit.

Note that the London Traffic Act, 1924, which checked free competition for passengers on London's roads, was of Conservative origin and passed in the interests of a strong financial group. It was certainly not the work of any wild-eyed reformers. Yet the Labour Party, from other motives, seems to think much as their political opponents on the prelimi-nary issue of competition or co-ordination. Herbert Morrison thus cogently sums up what free compe-tition, according to those who believe in it, *should* produce :

Good wages and employment conditions,
Adequate capital for new developments,
Adequate services everywhere,
Rock-bottom fares,
Safer, more pleasant and more comfortable rolling stock,
Reliable and speedy services.

He then proceeds to show (in *Socialization and Transport*) that these were just the things free competition was *not* producing in London transport, except in a limited way and at ruinous expenditure of capital which could not continue.

The minds of many responsible transport executives are moving forward on the lines of London's passenger transport achievements, and further passenger transport boards are being talked of. Herbert Morrison's party intend it, and frankly say so ; but whether his party presently gain power or not, the minds of senior passenger transport executives are working in a not dissimilar direction, and for self-protective and not idealistic reasons. Most people, indeed, want competition in everything but their own business !

Municipalities and Passenger Transport.

Whereas until fifteen years back transport regulation in Britain has been largely negative, laying stress on what undertakers might not do and seeking to protect the public from known or possible evils, since the War it has become more positive. Consider the Acts of 1921, 1930, 1933 and 1934. Under the 1930 Act the position of municipal passenger transport undertakings has been improved, for they now have greater opportunities to develop their systems and are put on equal terms with private operators. Very large operating units have emerged in road passenger transport, especially in urban and industrial districts. The small man still provides useful country services, but elsewhere the large road transport combines are with us, and are forcing municipalities to widen their outlook.

Dr. K. G. Fenelon recently computed that there were 137 municipal transport undertakings in the country, representing about £95,000,000 in capital expenditure, and owning about 12 per cent of the total public service vehicles in operation. Joint boards or committees to operate facilities of value to several adjacent municipalities are commonplace, but opinion is now tending toward the creation in some provincial areas of public passenger transport boards somewhat

44

like London's. A notable step is the formation of a joint company by the Keighley Corporation and the West Yorkshire Road Car Co. as from 1st October 1932, and somewhat similar arrangements were effected in August 1933 between the Hull Corporation and the East Yorkshire Motor Services Ltd., and in April 1934 between the York Corporation Transport and the West Yorkshire Road Car Co.

Then, since 1929, there were effected three separate co-ordination agreements between the municipalities of Halifax, Sheffield and Leeds and the railway companies ; while for some years a very ingenious agreement has existed between the Belfast Corporation and private bus companies working into the area. Several municipal transport systems have been bought or taken over by large motor-bus companies, e.g. those of Carlisle, Ayr, Kirkcaldy and Kilmarnock, and proposals of this type seem to be on the increase.

Agreements of these and even more fundamental kinds seem likely to increase in number so that Mr. Sydney Garcke is perhaps not premature in visualizing, as he did publicly in October 1934, as President of the Institute of Transport, a situation in which road passenger transport working in all large centres of population having been " grouped ", there would be left to independent operation only the smaller services running through sparsely populated districts. Since these alone would not be attractive to strong financial groups such as his own, he saw no alternative at that time to the creation of a National Passenger Transport Board to assume responsibility for road passenger transport as a whole.

Although these are possibilities, it has been intimated

on behalf of the railways that they would oppose any merging of municipal and other road transport interests that might result in the abstraction of "their traffic". Yet Mr. David R. Lamb, M.Inst.T., in October 1934 held the considered view that the only alternative to nationalization was the process so aptly described by Mr. Oliver Stanley when Minister of Transport as "eliminating the waste of competition while retaining its incentive". He was referring, be it noted, to nationalization of the "various means of transport".

Changing Basis of Railway Charges.

The rapid spread of the system of "agreed charges" for goods movement by rail means an increasing abandonment of the principles embodied in the railway classification and the Railways Act, 1921. As some experts put it, a planned, scientific system of railway rates built up throughout the railway era to serve the needs of the country and based largely upon "ability to pay"—a system possible only under a monopoly regime—is gradually giving way to what might be termed rule-of-thumb methods to meet the exigencies of day-to-day business.

Still, since the system is evidently giving an important section of the trading public the kind of railway charge it wants, and is guaranteeing to the railways an appreciable bulk of traffic, it is hard to see what real objection can be raised by people who have long urged the railways to "bestir themselves" and apply up-to-date commercial methods to their traffic getting and operating. Such people ought, indeed, to hail this as an encouraging sign, and ask for many more such. Others, whose business is not of a type conducing to an agreed charge, may have more ground for annoyance, although until recently, at any rate, the companies' exceptional rates policy also has been pretty practical, when skilled approach is made by trading concerns.

It might with some justice be argued that the system of agreed charges for traffic by rail, however questionable its genesis to lovers of fair play, is in fact providing the first real step towards a natural and more or less voluntary " division of function " between rail and road, in that the railways are contracting to deal with the whole of a given traffic and, in some cases, to deliver it to any part of the country, using rail and road for the purpose as best fits the case. An objector, however, might quote that witty civil servant Sir George Murray who, reporting on certain proposals affecting the Irish Office, minuted thus : " As far as I can make out, what he means by co-ordination is the subordination of everyone else to himself."

To-morrow's Possibilities.

Whether ideal solutions for the general internal transport problem will, in fact, be reached is, of course, open to discussion. While succeeding chapters are efforts to throw out suggestions for reforms helping toward an ideal, it would be foolish to ignore the immediate practical possibilities. The observed and now almost inevitable tendency towards large-scale administration in the pursuit of economies may or may not work out in the highest interests of the public. All depends on who gets the benefit of the economies secured, and at whose expense they have been made. The railway higher direction at least

47

have undergone a great mental change during the past seven or eight years. The old idea of a railway as a concern only interested in carriage over the rails has passed, and newer thought pictures it as a general transport undertaking, ready and able to undertake whatever form of transport best suits the needs of the case. Railways now are fully authorized to run such road or air services as they choose : they are large dock owners ; owners of many short sea trading vessels ; owners of some part of the canals of the country ; and are under agreements for rate maintenance with coastwise shipping and the canal carriers. The potency of the railway propaganda, political organization and discipline have been well proved during the past four years, and the public cannot ignore the likelihood that what the railway general staff really wants it stands a good chance of getting.

Against the obvious dangers of a return to railway monopoly run for private profit there is, in the goods transport field, one strong bulwark at present—the private or " ancillary " operator of road goods vehicles. Because of this very important element, the Road and Rail Traffic Act, 1933, is now seen to have failed to solve the problem of road competition ; so the railways are already turning their serious attention to engineering some further limitation of the ancillary operator, but treading very warily for fear of raising a howl too soon, and so losing further traffic now moving by rail. On the other hand, authoritative opinion among owners of ancillary vehicles holds that in a year or two the number of commercial vehicles belonging to private owners will very substantially be increased.

" Key " Position of the Ancillary User.

Ideally, this may be unfortunate, for it works against the pure theorist's hope for complete co-ordination of all forms of transport. But one must remember that transport in Britain is confined to a compact island with many opportunities of manipulating the venue of production (in many trades, anyhow, and in an increasing number as modern power supplies are widely distributed) so as to combat transport costs. Hence, a proper co-ordination of transport will only be possible when such guarantees of efficient service can be given as will reassure the private operator who now resents being interfered with.

Indeed, a very significant feature not yet heeded enough by the transport operators is the rapid rise to prominence of the new profession of industrial transport management, or the science of controlling on behalf of industry, the movement of all its raw materials from their place of production or raising, by diverse means and over many miles, to the factory ; the internal movement of goods at the factory ; and the subsequent assembly, packing and despatch of the finished products to every part of the home and often of overseas markets. From being a pet enthusiasm of a few prominent industrialists this study has become one of ever-growing complexity but vast potentialities for true economy. It is leading to increased ability to market goods at a reasonable price and in good condition, in areas not hitherto penetrable. Closely linked with, and making possible, the modern type of nation-wide advertising and national sales policies, industrial transport managers are now among the most senior and most trusted of business executives.

Efficient production is wasted unless followed by efficient distribution, and to keep abreast of the legal, administrative, political and labour moves while intelligently directing the freight movement policy of a busy manufacturing concern or group calls for unusual talents.

Industrial Transport Management—a New Profession.

Men possessing this high training and able to take the broad view-point are probably of superior national value to those who are expert merely in the daily operation of a single arm of transport; yet until recently their voice has been little heard in the nation's counsels. The steady rise of the Industrial Transport Association, the national professional association for this class of technician, bids fair to remedy this situation by the mobilizing or an informed and corporate industrial opinion on these matters, which may lead to the understanding and exposure of such moves in the transport game as are merely selfish, while moulding a constructive policy so wise as to convince even the most selfish transport official that the national interests may not be inconsistent with his own.

For surely these are the people who should be heard. Industry, as a whole, is the direct payer of freight transport bills of enormous dimensions. British traders send by rail even now about 250,000,000 tons of traffic each year, and it provides about two-thirds of the gross revenue of the railways. They own and use for their own purposes nearly 80 per cent of the 400,000 goods motor vehicles on the roads. Each year they send about 13,000,000 tons of goods by canal, and 38,000,000 tons by coaster. Whereas the proportion of our average transport costs to average cost of goods was in 1913 just 11 per cent, the figure had risen by 1929 to 17 per cent. An official figure that is fully comparable cannot be obtained for a later year, but since prices have dropped since 1929 while average transport rates have dropped but little, the ratio is likely to be still higher to-day. By way of providing data for a comparison, the railways convey the very complex traffic of the F. W. Woolworth Co. to any part of Britain for 4¼ per cent on its *cost* price, while, on the contrary, the cost of rail transport in merely assembling the raw materials needed to make one ton of pig iron is 20 per cent of the selling price of the pig iron, or 25 per cent on the cost price of the materials.

Relation of Transport to Commodity Prices.

That average transport charges continue at an abnormally high level is usually demonstrated by comparing average rail rates with average wholesale commodity prices. According to *The Statist*, wholesale prices when represented in 1913 as 100 rose to 165 in 1924, but fell again to 100 in 1931 and subsequently have been down to round about 90. The *Board of Trade Journal* gives another view of the position, showing that in February 1934, compared with 1913, prices of cereals, meat and fish and other foods were at the level of 104 per cent while other commodities represented 106 per cent of 1913. Railway rates at 160 per cent of 1913 prices present a wide disparity with other commodity prices—although it must be added that the railways have in recent years been granting exceptional rates for particular traffics which in some cases are much nearer the pre-War level. Though the coal and iron and steel industries

51

are fond of quoting that "60 per cent" figure as a feature of their sustained attack on railway charges, it has been authoritatively stated by a leading industrial transport manager in another industry (Mr. W. H. Gaunt, Distribution Manager, J. Lyons & Co., Ltd.) that he was paying some 20 per cent to 30 per cent above pre-War rates. The heavy trades, too, get a material relief from the operation of the Railway Freight Rebates scheme. Yet responsible coal trade leaders continue to draw adverse attention to the comparison of the 4¼ per cent on cost price of goods which now represents the flat transport charge, anywhere, for the miscellaneous traffics of an important chain store, as against the long-distance coal traffic which still bears a transport charge of over 100 per cent on the pit-head price of the coal.

Coal and mineral rates are important to the railways in that in 1913 the combined tonnage of these two classes of goods carried by rail reached 297 million tons out of a total railborne goods tonnage of 364 million—or 81 per cent. By 1933 the volume had shrunk to 210 million tons out of a total of 251 million tons, but the percentage had risen to 83½ per cent. Over the same period the total freight train mileage in Great Britain had dropped from 155 million to 122 million, and between 1922 and 1933 the average net load per train in tons had fluctuated from 128 to 122 tons. Hence, of course, the sting in the railways' reiterated complaint that road transport had stolen—unfairly stolen, they insisted—the cream of their traffic ; the smaller but still highly important quantities of goods in the higher classes on which the railways looked to secure high rates and a good contribution to net revenue. "If", said the railways in effect, " our cries are not heeded and our competitors' *unfair* advantages are not taken from them, we shall lose all our higher-class traffics and the whole weight of our heavy overheads will fall on the heavy mineral traffics, forcing us to raise our present low charges on them despite their essential character in the national economy and their dependance upon rail transport."

Railways and the Basic Industries.

Although, in the opinion of many prominent industrial transport managers, this plaint contained some most audacious assumptions, it proved effective, and the railways got their way. Indeed, the indispensability of railways to national well-being has been reaffirmed in the minds of many statesmen, and seems to have become a pivot in the argument of some groups of reformers ; though this surely is more of a compliment to skilled railway press bureau work than to the clear, unimpassioned thinking of the reformers in question.

Do Railways Subsidize Heavy Industries ?

So, before seeking in the next chapter to develop ideas of the fundamental changes needed in our outlook on the entire transport problem, we must examine a bit more closely this basic belief which colours so much of current transport thought—and, it may be, quite improperly colours it. In June 1933 it was brought out by a study of railway statistics officially published that, taking the unit generally used by the road haulier for a cost basis, i.e. the vehicle mile, the heavy industries pay the following price per truck (or vehicle) mile : coal and coke, 10·41*d.* per truck mile ; minerals in classes 1 to 6, 10·02d. per truck mile, against only 6·74d. per truck mile in classes 7 to 20 traffic which, of course, embraces the goods said to be taken from railways by road competition. In the pursuit of their own objectives the railways countered by stressing ton-mile figures as a better basis of comparison, and these for 1933 show thus :

	Receipts per ton-mile.
Coal and Coke	1·046 pence
Minerals and merchandise :	
Classes 1 to 6	1·021 ,,
Classes 7 to 21	2·097 ,,

But the trader critic replies, still using figures officially furnished by the railways themselves, that while the average length of haul of coal, coke, etc., is 42 miles, of other minerals, etc., in classes 1 to 6 is 64 miles, and of merchandise in classes 7 to 21 is 102 miles ; yet the average wagon load for coal and coke is 9·40 tons, for other minerals is 9·15 tons and for classes 7 to 21 merchandise is only 2·83 tons.

It is difficult to carry the argument much further without recourse to confidential data, but since 83 per cent of the railways' freight tonnage consists of coal and minerals (1 to 6) traffics which earns over 10d. per truck mile and causes the railways little trouble over collection, sorting and delivery, it is hard to see how, if that enormous part of the total freight tonnage is truly carried at unremunerative rates, the railways manage to continue to survive on their losses. Especially is this so since, by their own statement, the railways carry nearly 80 per cent of their goods tonnage at exceptional rates and have been forced to make the most drastic cuts in the upper classes which are most susceptible to road competition. It follows that much of the higher-class traffic retained by rail must be kept because, in the main, competitive rail rates have been quoted. These facts lead to a strong assumption that the heavy industries—bearing in mind that their traffic goes mainly in full wagon loads as against the small average load and the variety of work needed on multifarious small consignments—are, in fact, finding a big share of the railways' net revenue as well as their gross revenue to-day.

Even more striking support for this view comes from a memorandum handed the Minister of Transport on 26th January 1932 by the railway companies by way of comment on the final recommendations of the Royal Commission. Between 1923 and 1930, they showed, railway receipts for merchandise (except coal and minerals) had gone down from 100 per cent to 88·9 per cent, while in the same period railway expenditure had gone down from 100 per cent to 88·5 per cent. Now, if railway expenditure as a whole dropped from 100 to 88·5 per cent, and railway passenger receipts were further down, from 100 per cent to 82·7 per cent, while the receipts from the mineral classes dropped only to 90 per cent of 1923 (£72,000,000 against £80,000,000) there is at least strong presumptive proof that the carriage of mineral traffics by rail at least paid its way to the railways.

Railway accounts prevent the student from ascertaining with any greater degree of certainty whether the heavy traffics do or do not pay the railways, but the reticence of their senior officers when attempts are made to argue this point to a serious conclusion is in itself suspicious. The special importance of

55

attempts to resolve the point lies in the use the railways have made in political circles of the statement that they are nationally necessary because they carry the heavy mineral traffics at a low cost—and so anybody who interferes with them must be handcuffed.

CHAPTER FOUR

ACHIEVEMENT OF TRUE TRANSPORT CO-ORDINATION

IT has been said that the man who boasts that he is self-made relieves the Almighty of an awkward responsibility. It is quite certain that the man who claims to be an expert on any subject reveals the fact that he is past learning about it. In the recent past an exaggerated respect has been paid to the expert. The expert in finance has sought to keep secret the mysteries of his service because he thus retains the profits from them, but monetary reformers of various kinds are now finding him out and explaining frankly to the public, who will listen, how they have been duped. The job of business is to produce and distribute, yet in the past century those two processes have been forced apart, have failed in combination and co-ordination. So finance, which is not a primary function of economic life, but merely a useful servant, has become master in the house. The servant must again become but a servant. If industry considers what it is paying for credit, for banking, for insurance, for financing, for all the machinery which is concerned merely with moving the tokens about, and not with the work of the world, it will realize that throughout the past decade all over the world those parasitical services are the only ones which have been uniformly profitable.

Production and distribution have to be welded together once more, and finance has to do its job obediently and reasonably. That is becoming obvious to every unbiassed student, and it will be implied throughout all that follows here. In going on to consider what specific changes are immediately and more remotely necessary to weld our transport facilities to our real current and future needs, therefore, I decline to be unduly bothered by considerations of finance such as arise at once to the mind of he who considers himself a "practical man". (Unhappy phrase! The "practical man" seemingly would be content to grovel in the gutter, or scrape along with his limited and circumscribed life, for lack of imagination to see the really dazzling possibilities before us if we used our present vast knowledge aright!) To the man who asks at every turn, "Where's the money coming from", I have the simple and invariable reply: "From the same place as it comes from now, i.e. the printing press, or just an entry in a bank ledger." But instead of being issued in deliberately scarce volume and for the benefit of a private group, it will again be issued under Government authority and in such quantity as is required. That phrase means exactly what it says. Currency issued by a wisely run Government in a definite and calculated relationship to actual and potential production of goods and services (which alone are real wealth) is not inflation; and currency, which is but a convenient system of tokens which pass in fair exchange and convey no permanent title to levy toll on the results of future production, do not increase the now fantastically impossible load of book debts under which we are supposed to be labouring.

That being understood, what does one foresee and recommend? A preliminary answer will be given by quoting from a memorandum bearing on the co-ordination of road and rail transport, approved by the Industrial Transport Association and submitted in June 1935 to the Transport Group at the Congress of the International Chamber of Commerce in Paris. Within its intended limitations, this is the most masterly outline scheme which has yet been made public. For the first time a serious plan is put forward for the sympathetic welding together, for more efficient service to the community, of the main systems of internal transportation—road and rail. I outline the chief points of the argument :

True Co-ordination of Transport.

It appears to be generally assumed as a basis for consideration of this problem that the railways are essential to the well-being of the State in some particular way. While we can assent to this proposition broadly and while it can be said at the moment that in the case of some industries at any rate railways are absolutely necessary to them, we are not prepared to assent to a proposition that railways are essential if this proposition is to be in any sense exclusive, that is to say, if it implies a denial that air transport, canals, coastwise transport and road transport are just as essential to the national economy. (It must be understood that we are for the moment dealing with the matter from a purely commercial point of view : we are not introducing military considerations of any kind.) In regard to canals and coastwise transport, their place in the national life has been recognized for many years, although the treatment of the canals, either directly by the Railway Companies, or indirectly as a result of the growth of railways, should serve as a warning in considering the relations between rail and road transport.

59

In regard to road transport, perhaps its importance in the national life is best described in the words of a modern historian who cannot in any sense be regarded as partisan, namely Professor G. M. Trevelyan, O.M., who writes in *The Times* Jubilee Supplement of the 3rd May 1935 under the title " The Crown and the People—Twenty-five Momentous Years "—

" Motor transport has during the King's reign come into its own, transforming the manner of life of whole classes. The new method of conveying heavy goods is of immense value to commerce, industry and even agriculture, and it is a principal cause why the general standard of life has not fallen as a consequence of the war and the world dislocation of overseas trade."

Recent legislation in Great Britain and the resultant administrative-legal action which has followed have had a definite railway bias. This bias has been enormously influenced by the financial resources of the Railway Companies and by their organization for propaganda through the Press and other means of publicity and by means of evidence diligently collected and used in the Licensing Authorities' courts and elsewhere to contest applications.

It is a primary necessity for the objective consideration of the transport problem as a whole to get rid of any bias in favour of any particular form of transport and to secure that the opinions of those who are by the nature of their calling likely to be most free from any such bias, namely the users of transport, are allowed full expression and given full weight. As an Association of Industrial Transport Administrators, we represent in a highly concentrated manner the views of transport users generally.

In regard to the situation as we find it in Great Britain to-day, it must now be perfectly clear to everybody that the Road and Rail Traffic Act of 1933 has so far failed and is bound to fail to solve the problem presented by the competition of road transport with the railways, in so far as that solution in any way depends upon the elimination of the goods motor vehicle. It is true that the position of the public carrier is made exceedingly difficult and there can be no doubt that we have to anticipate a progressive elimination of the public carrier

(a) by the refusal of licences ;

(b) by taxation which is becoming punitive and

(c) to a limited extent, by effective competition in price.

Alongside this gradual elimination of the public carrier, however, there is a steady growth in the use of the goods motor vehicle by the private carrier, and while this remains as a useful safeguard against monopolistic charging either by Railway Companies or Public Carriers, on the other hand it clearly has the effect of preventing the Railways from maintaining their rates structure, which is really what the Railway Companies are endeavouring to achieve by the aid of legislation.

As long as the private carrier remains in being, a process (which, as road conditions become stabilized, will become more rapid) either of loss of traffic in the higher classes or of price reductions to enable the Railways to retain such traffic, is inevitable. The real problem, therefore, remains, despite legislation. It is only the direction of the challenge to the railway rates structure which has changed. The upshot of it all must be, however, a return almost precisely to the situation which gave rise to the legislation of 1933, namely that the Railways will ultimately be left with only that traffic which is not susceptible to road competition and in consequence, *ceteris paribus*, upon that traffic will be thrown the greater burden of railway overhead costs. It appears to be clear that this burden upon the basic industries of the country is one which they cannot alone carry, and, assuming the necessity both of the railways and of the basic industries from a national point of view, it is clear that unless support can be sought from the higher classes of traffic by reason of the elimination or further restriction of road transport or by a revaluation of capital with consequential scaling down of railway charges, the alternative to bankruptcy of the Railways would be some form of subsidy.

Is it, however, right to assume that the alternatives mentioned are the only courses open ? We think not ! We are of the opinion that although from the point of view of railway finance the position in 1933 was considered urgent it was a great mistake to have attempted any regulation of road transport (except that directed to public safety and decent conditions of labour) until the whole question of division of function had been scientifically investigated : here again the influence of the Railway Companies has been felt, and the results have been deplorable. They appear to have been inspired throughout with the *idée fixée* that road transport must necessarily be a competitor, and only in a very minor degree an auxiliary, of an effective national transport service. The Railway Companies have never really given to road transport the place it ought to have had in their operating economy. The efforts that have been made in certain directions to use the goods motor vehicle as an auxiliary to the railway services have not been satisfactory for the simple reason that, broadly speaking, the services provided have been mere duplicates of railway, or even railway-road, services already in existence. The Railway Companies do not seem to have considered, and do not, indeed, to-day readily accept, the idea that there are certain functions now performed by railway which could be more effectively and economically performed by road, and no real effort appears to have been made to study the far-reaching effects of a thoroughly reformed transport operating system which would make full use of the road vehicle within its economic sphere. The Railway Companies have only grudgingly made use of the motor vehicle in replacement of existing services. They have been too " metal-bound " in their ideas.

If these ideas are right, as from our experience we believe them to be, there can be no real quarrel about the nature of transport as a public service or about the consequences which follow from the acceptance of that point of view. Even the debatable theory of charging what the traffic will bear might be acceptable to those who to-day oppose it if the user of transport could be satisfied that the operations of a national transport system, by the appropriate use of each instrument in its proper place, were on the most economical basis. Upon this matter the user is far from satisfied to-day and the user of road transport in particular refuses to accept the idea that he must surrender its very real advantages, not only as to economy but as to service, in the interests of other classes of traders whose traffics it is said are not adaptable to road transport.

As to the practical solution of the problem in Great Britain, it appears that users will have to face an extension of national control and that it will be necessary to establish a Transport Board, superior to railways and other forms of transport, which will include representatives of the carrying interests and of users in such proportions as to ensure that no particular carrying interest is able to turn the scale to its own advantage against the national interest. (In this connection some special provision would have to be made for the representation of ancillary user interests in order to ensure that those special services which must necessarily be controlled by the trader himself are preserved.) It is idle to pretend that the activities of the Licensing Authorities will result in an accurate survey of the transport situation which will enable any transport advisory council scientifically to determine the functions of the respective branches of transport. This matter can only be studied in the greatest detail, area by area, over a period of years and will call for not only the highest scientific and administrative knowledge of the transport industry itself but also for the best commercial knowledge of so manipulating the movement of goods as to ensure that appropriate use may be made of each branch of transport, with the maximum efficiency and economy.

It must be obvious that the waste of national effort in transport to-day is enormous, and we are perfectly satisfied that a proper co-ordination, based upon a scientific rather than a politico-financial division of function, combined with periodical capital adjustments which will relieve the transport user of the burden of ineffectively-used plant or equipment, will so reduce the general cost of transport as to ensure that there is little room for differences of opinion amongst the users of all forms of transport, as to the nature of the rates structure necessary to place the transport system as a whole upon a sound economic foundation.

This memorandum was one of several very important outcomes of a National Conference of the Industrial Transport Association on the general theme of " Transport Planning in the New Age ", and at that conference, summarizing papers containing much valuable detailed material, Mr. Dudley A. Elwes, F.I.T.A., the Indoor Traffic Manager to Lever Brothers Ltd. at Port Sunlight, saw in them all two main and proven assumptions :

(1) That certain existing transport conditions are so obviously deficient and faulty that an instant remedy can and should be found. And that other weighty difficulties call for the highest genius in the organization to be applied towards their solution.

(2) That no modern transport manager can afford to think in terms of only one or two forms of transport, and that the operations of carriage by rail, road, coastwise, canal and air at once connect his local work with national outlook and policy. In other words, he must be balancing continually his own traffic costs and advantages in these five arms of transport, against the national interests from which they are now inseparable.

What is an Adequate Service ?

Next one must dismiss—or, at least, place in correct perspective—the normal comments as to " adequacy " and " duplication " of services. To try to define adequacy in any quantitative manner would involve so many contentious points that the main argument would be obscured ; so one must try to do it in a general way sufficiently indicative of the line of reasoning. Is our general standard of life adequate for *us*, and for everybody ? Until it is, how does the " adequate " argument arise ? Are we content with poverty amidst plenty ? In view of observed human nature—ambition, acquisitiveness, discontent, etc.—shall we ever within human range of vision reach a state of true adequacy in our transport supply ? New facilities beget new demands if not always new needs. The progressive shortening of the hours of labour should mean greatly increased passenger travel, in any case, and a more scientific control of currency tokens should rapidly increase the effective demand for the wealth we can produce—beside leading us on to a further wide range of queries. Is the present location of industry permanent ? Obviously not, but in what particular ways or directions will it change ? What results will follow the effective completion of the electric power grid as the public is educated to demand and use its facilities ? What new changes in methods of capturing or creating power and transmitting it where needed, may shortly be upon us ? How far will slum clearance schemes fructify in the immediate future, and how quickly will garden city enthusiasts and others working for decentralization of population force the slow-witted to see the sense of what has so far been achieved ? Without definite answers to these, and many such points—and who dare give them ?—how can it be said whether or at what point our transport facilities are " adequate " ?

Plan for Abundance.

We must plan, therefore, for abundance, and get our thoughts right above the plane of the " work fund " fallacy which still dominates so many of us.

The tragedy is that so many people holding responsible positions in control of affairs to-day do not rise above this stupid fallacy. Hence most of the agitation, financial group rivalry, biassed propaganda, and so forth anent what for so long has been termed the " rail versus road " controversy. The memorandum above quoted points the way out of this morass of unhelpful friction, but what a tragedy it is that such ideas must still be forced upon most of us. The game we're playing is not a purely local " friendly " club fixture of " road versus rail " which we can watch with a sportsman's keen appreciation of the subtle tactics employed by the contestants, but a really grim struggle of " Britain and her people versus poverty, unemployment, and callous because deliberately continued shortage for the masses ".

We have to plan for prosperity ; and that means the prosperity of the people as a whole as well as that of transport operators of all sorts. The Ministry's Transport Advisory Council *may* be about to make a strikingly far-sighted contribution to thought thereon : but one doubts it, for members of that body likely to think constructively and nationally will probably be outvoted every time by those speaking for one or other transport operating interest ; and each member of the latter type is likely to continue actively finding ways of checking the others.

Planning, then, for plenty and not shortage, the aim should be to encourage and develop all our transport

66

assets, with due regard to common sense, providing only that the volume of transport service available keeps reasonable pace with the nation's valid needs. By instinct, I somewhat fear a National Transport Board having full compulsory powers, merely because I fear the effects of the application of a current public service mentality to these vast questions. Yet if plans can be devised—and why should they not be?—by which the incentives of ambition on one side and the true urge to service which does inspire some competent persons, on the other, can have adequate rein, I may well take further courage. If, also, general progress is so marked that these public servants are above the insidious personal temptations probable during the transition period, my main doubts vanish and I agree that the immense public benefits from ousting the financial parasites will outweigh a lingering doubt even I may have of the need for large scale enterprise in these matters.

So that by a somewhat different route I reach a goal very like that envisaged by :

With important distinctions : the financial controllers of our railways and road passenger transport concerns.

The Industrial Transport Association, consisting of the responsible officers trained to view these questions from a broad standpoint as they affect British industry.

The Labour Party.

The Co-operative Party.

Herbert Morrison would Safeguard Railways.

But in taking the next step I find myself gravely at variance with the Labour Party—at any rate in

67

the manner of stating their policy often adopted by Mr. Herbert Morrison. Without doubt the Labour Party, not yet having openly espoused the truth of Social Credit, feels itself bound to shape a policy which ensures the friendly votes of many thousands of keen N.U.R. and R.C.A. men. So Mr. Morrison, who has had such opportunities for statistical study of the subject that one thinks many times before differing from him, after premising the creation of the National Transport Board, says that its main line of policy would have to be the maintenance of the railways, almost at all costs, since they are essential to the heavy industries and so to the nation. Railways, runs his argument, must have good and constant loads to carry, if their heavy overheads are to be properly shared so that they can charge economic rates on all their traffic while reasonably remunerating all concerned. This being so, all other internal transport systems must subserve the railways' interests. Such traffic as railways can handle efficiently must be handed to them. As their volume of carryings increases, so must their general efficiency increase—very greatly increase—and hence not only can they soon supersede the public road carriers for most purposes but at the right time they can fairly expect to handle, to his advantage and their own, most of the "ancillary users'" traffic, i.e. that great tonnage of goods now conveyed by traders in their own lorries.

Like the Industrial Transport Association, I deny the view that railways are essential if it means that they are *more* essential than road transport, canals, coastwise shipping and air transport. The maintenance of *all*, with due economy and in their proper sphere, and without fear of vested interests, is what is needed—even of the vested interest created in the valuable support of several hundred thousand railwaymen's votes.

Overhaul of Railway Capital.

By every line of approach, one of the main things needed appears to be an overhaul of railway capital. The railways have been bemoaning that road competition and trade depression have prejudiced them. They put those factors in reverse order when it suits them. In truth, heavy burdens of unremunerative capital, while making no difference to the actual earnings of the railways to-day, make a big difference to the apparent results as seen by shareholders and investors. In early railway days many companies were promoted. As the Royal Commission of 1928–31 said : these promotions were " good, bad and indifferent " and many were initiated and carried out in an atmosphere of ignorance and prejudice which has left a lasting mark on railway economics. High prices were paid for land to buy off the opposition of influential landowners and meet claims for compensation in respect of depreciation, real or fancied, to estates. Thus the land acquired was often not that best situated for rail transport, and the capital of the companies was grossly inflated. " With such excessive capital expenditure ", declares the Royal Commission, " two things only, each undesirable, are possible : either the capital remains unremunerative or remuneration must come from excessive charges to the user."

Much wasteful outlay arose in needless competition between rival companies and we feel its costly effects still. When the Railways Act, 1921, combined over 120 separate undertakings into four, it merely recognized the futility of continuing an absurd state of inefficient organization. But the 1921 Act has proved inadequate, though it was a useful step. Voluntarily the railway companies have entered into pooling arrangements which practically eliminate competition among themselves. Except that the railway staffs are gaining experience, at our expense, in larger-scale operation, the present is a not very satisfactory half-way house giving the public many of the alleged disadvantages and few of the benefits of unification of railways. Unification at least is bound to follow soon. All the theoretic arguments for grouping apply with equal force to unification. But capital overhaul is equally necessary, and in the negotiations for the taking over, however accomplished, the chance should be taken to deal fairly with the shareholders by exchanging their share certificates for new stock at a basis computed on the average market prices of their holdings over a period. The "widows and orphans" argument is so familiar that it carries little conviction. Although railway shareholders deserve justice, they are not entitled to preferential treatment over all other industrial shareholders. By doing them justice in the transfer of their properties to unified control with commercially-minded direction, the first big load would be eased. This, and the many more technical points on which the railway system must be thoroughly overhauled, are more fully dealt with in due course.

Reverse the Tendency to Strait-Jacket Transport.

Were railways thus rationalized, perhaps their political pressure would be eased, and other services would need to be freed of questionable limitations put upon them in recent years in the railways' interests. The detailed applications of the principle to other transport services follow in later chapters, but it will at once be clear that it is proposed to reverse all tendencies to strait-jacket the providers of transport service, requiring indeed of them the provision of expanding and increasingly efficient services for goods and passengers to keep step with all other sorts of progress. The whole angle of approach to these points must change. The aim must be to get goods and people where desired, in safety and comfort, and *not* just to secure the longest length of haul for the originating line, as now. Needs will have to be ascertained, and systems adjusted to meet those needs adequately and in fitting manner : no longer shall we force our needs within the cramped limits of current systems.

Encourage Cultural Travel.

Cultural travel, for instance, should be deliberately encouraged. Passenger transport, indeed, should come to be regarded as almost entirely a cultural service. It should be cut to the very minimum so far as concerns travelling to and from work, and for the carrying of goods, since the process of scientific planning of production and distribution will continue and, though the volume of goods in effective demand would be much vaster than to-day, modern power would make possible a drastic re-location of industry, and railhead warehouses (and their counterparts for road, sea and air), with a radial distribution, would help retain the freight tonnage within manageable dimensions, especially during the transition stage.

Passenger transport, on the contrary, would be encouraged in every way and would be much cheaper, if not free, to the individual. Some authorities do

71

not go that far, but see a case for further extensions of differential charging : necessary transport such as travelling to and from work and the transport of food might be charged on a lower scale entirely than travelling to attend race meetings or the dogs. We already have workmen's tickets ; these facilities may be extended more than at present to children going to school or travelling undertaken for educational purposes, or on convalescence.

There may well be difficulties about this kind of moral censorship of one's varied activities by the grant of differential fares. In my own view, though the present practice in this regard will need to be continued and intensified during the next few years, as we throw overboard the economics of shortage and our lives are re-adjusted, most of the need for differential charging will disappear and a generally simplified and quite low level of rates will suffice.

Next Steps with Road Transport.

While long-distance road passenger transport might very simply be taken over and operated through the National Transport Board, it might be wise to encourage local authorities to retain some interest at least in organizing the local transport of urban communities—subject, of course, to their coming to suitable co-ordinating agreements, as seemed necessary, with nationally operated transport entering their borders.

Goods transport by road is far more difficult, but here also private enterprise, backed by the financiers in some cases, may be getting the situation into reasonably good shape for taking over. Leaders of the larger goods haulage concerns were reasonably content with the railway agitation which produced goods vehicle licensing, for presumably they foresaw it would help to eliminate the small man and strengthen their own hold on the traffic. They are building up frequent and reliable road goods services and are shouldering responsibility for " smalls ", at one time their great stumbling block. A little more rationalizing, leading to better use of overhead organizations, fuller loads and less empty mileage, is undoubtedly possible and likely. There will not be too much difficulty about the Board taking over the long-distance goods haulage concerns that remain, and probably in dealing similarly with the larger local public carriers, but at first it might be well to leave the smaller local and village carriers to operate under their licences. Ancillary users' vehicles remain the difficulty from the organizers' point of view, while also being the traders' main safeguard in this transitional stage.

Though canals may have a limited future there is no reason why the example of the Grand Union Canal Co. should not be followed, and such improvements be effected that, coupled with the re-location of industry, they may become again a reasonably prosperous and nationally valuable asset. The speed of canal transit is being improved, but for its most readily apparent uses great speed is not really vital. So long as a reasonable continuity of flow of fuels and raw materials by canal is maintained, and craft arrive daily carrying their required quota, industry would be quite happy.

Ports a Tough Problem.

The problem of the organized development of aircraft services is approached later, and attention is also given to the national planning of ports and of

73

coastwise shipping. In the case of ports it seems likely that especially vigorous opposition may be encountered, since some of the port trust chairmen have so confused the function of supervisor of port progress with one of purely personal honour carrying dictatorial powers that *anno domini* alone may shift them. However, one hopes over-riding powers may be given to the National Transport Board to point out unmistakably to any reactionary port authority that it is industry's money they are spending and that citizens far beyond their local municipal boundaries are directly affected by their actions and attitude. There is much surface plausibility in the arguments that can be advanced by the autonomous and municipal trusts for exclusion from any scheme of national supervision, so that this case must be more fully argued when the problems of the ports fall to be dealt with later.

Ocean shipping, too, will come in for some attention. It might have had a *prima facie* case for remaining independent, but it has graciously forfeited any such right by coming to the Government for aid while at the same time so loudly declaiming against foreign subsidies to merchant shipping.

CHAPTER FIVE

TRANSPORT AND GENERAL ECONOMIC PLANNING

DURING the Great War the Ministry of Reconstruction studied the unsatisfactory condition of the electric supply industry. The Haldane report of 1917 was the forerunner of the 1919 Act which created the Electricity Commission to inquire into the possibility of reorganization on a national scale. Eventually action was taken on the Weir Report of 1925, and the Central Electricity Board was created to carry out the work of concentrating energy in certain power stations : to erect a high tension main transmission system to inter-connect these stations and build up the existing regional systems into a national " Grid ". The Board does not own the means of production as represented by generating stations, and has no control over distribution. Its function is to make production more efficient through concentration and co-ordination, the instrument of co-ordination being met by the main transmission system. The Board makes its own arrangements with supply undertakings, fixes its own tariffs and controls absolutely its own administrative organization. Thus, it is of a new type which combines public control with a good deal of independence in operation and the criteria of success are those which obtain in industry generally.

The Central Electricity Board is thus a very significant unit in the trend towards economic planning. To quote *The Times*:

> A study of the similarity between the Government's agricultural policy and the Grid organization of the C.E.B. may provide a clue as to what the new model for industry ought to be. This is an interesting association of an experiment with a remarkable success. Freedom in efficient production is linked with freedom in efficient distribution by a co-ordinating board, exercising initiative and stimulating enterprise and development.

With the completion of the Grid a new and vital factor comes into our planning problem. When the electricians climbed down from the steel tower at Fordingbridge near the New Forest at 11 a.m. on September 5th, 1933, it was an occasion worthy of epic celebration.

The Grid is now beginning to perform an important national economic service in accelerating the modernization of industry consequent upon the reduction of power costs. Through the pooling of energy supplies, the Grid makes it certain that the economies which hitherto have been confined to the very large centres of demand will be spread over the whole country. This must lead to better territorial distribution of economic activity and opening up of economically undeveloped areas accompanying, or perhaps resulting from, industrialization of agriculture and the closer adjustment of revenue to capital expenditure involved in distribution. Thus, an electricity supply undertaking, in considering the possibility of extending a distribution system over a sparsely populated territory with few industrial potentialities, no longer has to bring into calculation the capital involved in providing for the generation of electricity, for it can get electricity from the Grid at a price common to an entire area and so can base its estimates of revenue, costs and capital charges on distribution only. Apart from the industrial power load, rural and domestic supply at increasingly cheaper rates will be the tendency, and it should be most marked in rural areas which hitherto have had to pay the highest rates. Already authorized undertakings are developing their outlying areas and have penetrated rural districts, and with special success in Dumfries and Wigton, Mid Cumberland, the district around Northallerton and Malton, Mid Lincs, East Anglia, Eastbourne, Ringmer, Wiltshire, Devon and North Somerset.

Electrification Revolutionizing Agriculture.

Already one can see how electrification can be an important factor in converting agriculture into an industry able to use up its own products and reduce waste to vanishing point. A good example is provided by a concern in the Midlands which farms over 6,000 acres, feeding into a jam-making and canning factory on the one side of the territory devoted to a certain range of products, and a canning factory on the other devoted almost entirely to vegetables. This industrial concern has solved the problem of combining mixed farming and fruit-farming with the conversion of farm products into more valuable marketable products, and the power used for doing so is almost entirely electricity supplied by a company wholly dependent on the Grid.

Three important changes are taking place—the revival of economic activity in the rural township or village, the industrialization of agriculture, and the

77

location of new industries destined to meet the requirements of the consumer in those self-contained power
areas. Thus the new power network makes more
feasible than heretofore a decentralization of industries,
a better economic balance between agriculture and
manufacture, and some revival of local activities.
Equally, the Grid removes one of the grave disabilities
under which the national manufacturer seeking better
sites has hitherto suffered. That important class of
man can now re-survey his problem and build a
modern, efficient factory on low-priced land with
fitting modern access for one or more modes of transport, choosing as his venue of production a site which
is well balanced as between the sources of raw materials
and the general area in which the finished products
are distributed. Industry, in so moving, tends to
move also some at least of its operatives of all grades,
so that the need for modern and diversified housing
accommodation in new areas becomes imperative.
Pressure of public opinion and the need for a national
manufacturer to be sensitive to consumer goodwill are
likely to ensure, if nothing else could, that the new
constructions will be in reasonable accord with modern
standards, so that an impetus is given toward a higher
standard of housing and town planning comfort.

Beginnings of True Industrial Planning.

Indeed, every aspect of the general problem of
economic planning, i.e. increasing production and
facilitating wise distribution of the abundance now
thrust upon us by nature and science, must be closely
related to the various regional planning schemes and
town planning schemes already put forward by advisory
committees in the different areas. Such schemes

already cover a wide range, but, unhappily, they prohibit rather than construct. They survey areas which may or may not be put to industrial use. They do not yet show how local authorities, vitally interested in the proper development of their areas, can be directed to the end that proper development *will* take place.

Much more specific work must yet be done by regional planning and town planning experts, and until it is done our next steps are bound to be somewhat conjectural ; or rather, if they are to be sane, they must be conjectural. Unless they take real heed of these broader possibilities of change, they will continue as in the past to permit the indiscriminate development of transport services each with an eye merely to the " main chance ".

But surely before long the people's common sense will force regional planning to establish a balance between utilization of local resources and social needs of the population. Why should a great capital like Glasgow, one of the wealthiest in the world, have the worst housing conditions? In a striking sentence Fenner Brockway in " Hungry England " contrasts the appalling sanitary arrangements in the Bridgeton district of Glasgow with the unemployment in the greatest sanitary engineering works in the world, at Paisley and Barrhead. The inference here is that the increment in wealth and capital which is a condition of progress under capitalism in any area should mainly be allocated to the public service of that area. Capital should not have full liberty of movement, if such liberty means economic deprivation in any area and excessive over-building in another area.

Planning, geographically, has steadily enlarged its

79

scale during the past twenty-five years. What started as the Garden City movement has grown into the Town Planning movement and further increased its scope until regional and national planning are being seriously envisaged. An important aim of planning is to reduce the friction of transport and one of its most difficult problems is to facilitate co-ordination of the various means of transport. The advent of aviation has added a new problem which opens up a grand avenue of hope if only we have the sense to plan for it in time. Otherwise it will make confusion worse confounded.

Regional Planning's Early Achievements.

One result of regional planning has already been the reservation for agricultural purposes of considerable areas, and also private open space reservations where scenic or other amenities are worth preserving. Many miles of new streets, widened streets, and better building lines have been achieved and the zoning of specific areas for industry, residence, shopping, and civic administration has made material headway. Isolated results of great importance may be noted. Valuable industrial areas alongside tidal water and railways have been preserved for industrial use, and steps are being taken to secure a measure of conceal-ment of the industrial buildings by a belt of trees. Areas adjacent to canals have been reserved for wharf-age purposes. In this type of activity economies often arise to all concerned. Opening of new routes has meant saving of time and money in transport and travel. Statutory undertakers can foresee future needs and arrange their services accordingly.

Railways should begin to get attention from the

regional planner from their amenity as well as their economic viewpoint. It was most unfortunate that, as the extraordinary advance in speed of travel came to be appreciated, it was accepted with such fervour that operations were permitted that are now seen to have been very detrimental. Embankments and viaducts have wrecked the orderly plan of many areas, and made good planning impossible in many suburbs. Public open spaces, such as the Surrey commons, were allowed to be cut up by railway routes without compensation, and it is only quite lately that railway proposals have been opposed on the ground of their affecting the amenities. It is easy to see that, in the experimental stage, encouragement was needed and large concessions were perhaps not unreasonable, but this attitude seems to have been maintained long after it should have been abandoned.

Railways and Aesthetic Considerations.

Suggestions have been made that in towns no railways should be allowed to be carried at a level above that of the streets, and though this would remove the very ugly bridges our engineers seem to delight in, overhead lines might surely be allowed if skilfully planned and carried out with some thought of artistic effect. Moreover, while railways survive we ought not to ignore the railway traveller in favour of his brother in the street—though the railways themselves have always done so. On a question of outlook, the man in the train has some claims, but hitherto both parties have been to blame ; railways for disregarding the harm they were doing to districts they ran through, and builders for disregarding the railways and taking no care that a presentable face was made to them.

Possibly in the future the backs of houses will present no worse appearance than the fronts, and they will all be placed far enough from the trains to allow of some planting and gardening between.

The Shape of Cities.

Glancing to-day at most cities, we find administrative and business premises centrally placed, with industry and commerce stretching out in one or more directions, and residential quarters in others ; commerce seeking a place where rail and water transport is available (or more recently, an arterial road) and residences the areas most attractive and accessible. This broad generalization is, however, much qualified in practice, for industry and commerce draw into the interstices of their quarters a good deal of housing for the workers employed ; while industries less dependent on transport distribute themselves among the residential districts. Thus, instead of a defined scheme of production, transport and distribution, we see only a blurred travesty of such. Much of this has doubtless been due to lack of adequate provision for transport and transit, which, instead of being planned in advance, has been improvised to cope with the demands after these have made themselves evident. So, it is not often possible to do all that might be desired for our existing towns, which must be regarded as illustrations of certain forms of failure to be avoided in planning for the future.

In laying down zoning principles we must bear in mind that it is impossible to predict the future with absolute accuracy, and so alternatives in utilization should be admitted to the maximum extent compatible with the validity of the scheme as a whole. Hence in some parts more than one type of occupancy may be allowed. Even this is a big step forward, for the latitude formerly customary has left us with an awful confusion in our towns, as detrimental from the health and amenity aspects as from the standpoint of economics. Large areas of low-lying ground are covered with houses, and the smoke from factories on the neighbouring riverside streams over them. Elsewhere big industrial concerns have had to abandon their works because the ordinary road communications on which they relied have become obsolete.

Chief Considerations in Zoning.

In the planning reports already issued, close study has been given to the circumstances of each case before zoning regulations have been laid down. Some of the general principles observed may be outlined. Probably the business section of a city will stay at the centre. In a growing community it also must expand, and is likely to do so by increased height of buildings and by extending into adjacent areas occupied by residences or factories—a course not without advantages, for it is usually the older and obsolescent houses and factories that go. But the growing height of buildings will intensify the traffic—as New York has found to its cost—and so proportionate street widenings and very much increased passenger transport either under or on the ground are essential.

General industrial and commercial undertakings, which used to depend upon rail and water access, now tend also to demand good road access. Of course there is a steady move out of urban areas, but where there is navigable water its frontage will be sought after, both because of the comparative cheapness of water transport and because water is useful in connection with some manufacturing operations. Direct rail access also remains important for many concerns, and this is more easily organized on the flat ground, undesirable for housing purposes, to be found at the lower levels. So we have good general guidance, and can reserve the higher ground for residential purposes. If this is irregular in conformation, it will lend interest to the lay-out, and any land too steep to be built on with economy can be planted and developed as strips of park with pleasant walks, breaking up the monotony of large areas of building.

While in the cities everything else favours a clear separation between large industrial undertakings and the homes of their workers it must be remembered that if these groups be segregated, convenient and economical means of transit must be provided between them.

Control of Site Selection.

Until recently industrial concerns have had a fairly free hand as regards site selection, and while the positions chosen have often been injurious to the general community, they have usually been taken up with a view to advantages of one kind or another in relation to the operations contemplated. Since restriction of this freedom is inherent in regional planning, it is the more necessary to ensure that the areas allocated to industrial purposes shall be such as will enable these to be economically and profitably carried on, and that they can be apportioned in sites suitable for the classes of industry likely to be in demand.

This question can be approached with greater

84

clarity if the operations of industry and commerce be sub-divided into their constituent parts. Beginning with " raw material ", we have first the processes by which this is gathered or won, next its transport to, and reception at, the place where it is to be worked on, and possibly a period of storage there ; then the manufacturing processes, which may be the complete series to the finished article, or only a partial one preparing it to be passed on elsewhere, as in the case of a smelting works or a tannery ; then another period of storage and the distribution of the finished or partially finished article.

The appropriate relationships between all these operations demand thought, and where the assembling of various constituents or the provision of power comes into the account, it is often a question of great nicety in economics as to which factors are to dictate the location, whether ore is to be brought to coal or coal to ore, whether material to power or power to material. Successful prosecution of an industry has in the past depended not alone on facilities in respect of material nor as to ultimate markets, but on a balancing of both these factors, together with others, such as competing demands for labour, climatic suitability, etc. Hence the wisdom of planning industrial areas with as much flexibility as possible is recognized, so that provision may be made alternatively for smaller or larger undertakings, with lines of communication spaced out so that supplementary ones can be formed when the actual allocations are defined. Valuable water frontages should be economized so far as the character of the industries permits, and a good depth given to sites so situated, as in most cases branches of the work will not demand proximity to the water-front. Where the principal means of transport is by rail, the planning requires care in providing that sidings can be carried to the necessary points without undue waste in arranging curves. It is often best to align the sites obliquely to the railway route, so that the sidings may be run into them without too long a curve.

Special Zoning Problems at Ports.

Nowhere does the difference between good and bad planning so seriously affect the economic welfare as in the large commercial seaport, and it is here, owing to the rapid advance made in marine construction of recent years, that the greatest activity has been needed, to keep pace with the progress made. Many of our wharves and docks and their plant are obsolete both in scale and efficiency, and only very occasionally does their position render possible a solution that has proved sound in Liverpool, where several important office buildings have taken the place of one of the older docks. In London and some other places where the demand for an enlarged business centre is pressing, the same plan might be followed and extensions secured by abandoning the older docks, which are outgrown by to-day's shipping.

New dock planning depends on the class of trade and the size of vessel expected, the former determining the proportion of land area and the arrangement of the warehouses and shore communications, the latter the scale and design of the water areas. Clearly, the arrangements of the docks or wharfage is the most vital factor in the extension of a sea port, and efficient planning is impossible where this is treated carelessly or excluded from the general plan. The physiography of the site must influence largely all such planning,

86

helping to decide the basic question of " closed docks " or " open quays " or both. The two main variations of the open port give wharfage parallel to the shore, running the length of natural or artificial frontages, and, alternatively, a series of piers at right angles to the shore line. The first and older type is general in Europe and the second in America. Enclosed docks show similar variations ; in some the ships lie along the walls of the dock, and in others along jetties projecting from these walls. Formerly these jetties were placed at right angles to the wall, but an oblique position now proves more convenient for manœuvring ships, and gives more economy in the space taken up by railway sidings, just as in the oblique arrangements for factory sites.

Port Accommodation Features.

The special planning for port areas in the way of sorting sheds, storage, elevators, bonded warehouses and the like, need not be gone into here, but when the general lines of development have been settled and the necessary areas allocated, there remain the lines of access and communication both for the service of the port and for the sites in proximity to it. Since these sites will usually be of eminent value for industrial and commercial purposes, it is vital that, apart from the lines of road and rail communication between the port and its hinterland, there should also be, distinct from these as far as possible, transport links between the port and the surrounding sites occupied by industrial concerns. Formerly these were often served by inland canals running from the port. Railways now do much of the work, but these two might be supplemented by overhead transport, such as travellers or rope lines serving large works, near the port area. The bold and successful experiment of the Tilmanstone Ropeway connecting the colliery of that name with the Admiralty Pier at Dover is an encouraging example of what might be done.

As port organizations usually occupy low-lying ground, one hopes that, avoiding the bad practices of the past, we shall only find sites suitable for housing the workers at some distance, leaving the zone close around the port for appropriate industries, and arranging to bring the employees from the more desirable places in the district by a rapid transit service. Such form of organization is urgently needed both north and south of the Thames estuary, and also along the Tyne, the Tees, the Dee and several other rivers. The lay-out of communications for passenger transit demands careful study, not only to avoid conflicting with those for goods transport, but also to give routes that may be economical in working while yet providing varied routes according to workers' needs. The housing of the workers in each group of docks and works at the nearest appropriate point, with a direct connection between these, sounds a simple solution, but it is inadequate, for members of a family may work at different places, and workers may change their place of employment without wishing to move their homes. A compromise between direct routes and a circulating one may best fit the circumstances.

Survey of National Resources Needed.

When we review the country's industrial possibilities as a whole, it is clear that the exploitation of these exercises a preponderating influence on the problems of development, and determines the general distribution of the people and, to a great extent, the commercial organization. A survey of natural resources, present and possible, is therefore the fundamental preparation for all future planning, both around the existing towns, and of the whole country, so that all likely expansions of industrial activities will be adequately provided for as parts of a general scheme.

Could all trade and industry be so grouped and planned that transport was cut down to vanishing-point, it would reduce a factor of cost to industry, modify the standard of living and leave the community with only the need for provision of roads for minor distribution and sightseeing purposes, in a rather drab and limited world. Conversely, if trade and industry were allowed to continue being indiscriminately and sporadically located, the volume of transport to carry goods about would thrive to the detriment of factory and production costs.

The truth is that, while transport is to some minds a necessary evil, its capabilities are so great and far reaching that it affords resources which, on the whole, benefit trade and mankind. While it is a relatively cheap service in many trades, a body of expert opinion favours fixing the production point at the least expensive assembly point, in transport cost of the raw material. The experience of Messrs. Crosse & Blackwell outstandingly demonstrated this theory. Though there is room for argument, majority opinion holds that a location which secures economical transport costs before and in production should be chosen, and the resultant saving can be used to effect a larger and more frequent service of distribution. This is actually taking place to-day in important products. It is possible to make great headway through the savings in mass production at the right centre, which more than compensate for the extra cost of more frequent and far-ranging distribution.

Housing Workers in Relation to their Employment.

As Mr. W. H. Gaunt pointed out to a recent Town and Country Planning Conference, one cannot argue on quite the same lines for passenger traffic. While goods transport is a direct charge on production, the worker pays his own carriage to and from work. If all tastes and ambitions were alike, and employment constant, the workers might need nothing further than a domicile within walking or cycling distance of their work ; but mass production involves a mass of labour, and with shorter hours the bulk of male workers are indifferent to a few miles travelling to and from work. They know that trade fluctuations may create the need for ready transfer to other employment, and the cost of quick transit can be more than met by cheaper house rents and the material advantages of good health and better recreation for leisure. Women are not such willing travellers but their work is usually only for a few years of their life.

So the objective is clearly to plan residential and industrial zones so that by a reasonable tax in transport cost and time, modern industry fits in with modern standards of a working existence, while giving to heavy raw goods transit problems the main preference.

To the extent that British industry depends upon imported raw materials, these thoughts point clearly to the continued movement of some manufacturing industries to the industrial zones in or near the principal ports. But, bracketing other considerations with our present thoughts, this may not mean the development of one single enormous plant of a given kind at one wisely chosen point, from which demand all over the country will be met. According to the volume of the article produced and its degree of " household necessity " the more modern policy of splitting production between three or four, or even more, well-planned works and delivering from each within a defined radius is taking shape. This point will be pursued later.

Thus regional planners, while rendering movement less necessary by placing things in better and more convenient relation to each other, will certainly not eliminate the need for transport service, but they will succeed in cutting out many of the occasions of waste and enabling every outlay in providing and in operating transport to be of positive productive or amenity value to the community.

Special Industrial Areas.

The development of special industrial areas such as Trafford Park and Slough has some manifest advantages, but it has probably received a check from two things—the vaster possibilities of road transport, and of the electricity Grid : beside two less important ones—fuel oil supply and the railway container system. These factors have already brought a very large amount of suburban and country land within the possibility of factory and industrial use and have had good effects in providing both passenger and goods transport with more balanced loads. In many directions now the morning outwards flow of labour from London and the evening inwards flow do much to remove the unremunerative peak load feature ; while the outwards flow of raw materials from the docks balances the inwards flow of finished products to the Metropolis, with no less facility for provincial distribution than if the factory were in the heart of London.

But the tendency to decentralize contains, from a transport viewpoint, one big danger. The indiscriminate or " sprawling " industrial development, the placing of a single factory miles from any others just because road transport and electricity make a cheap site possible, may well be definitely anti-social ; for it may force statutory undertakers to incur uneconomic outlay, lead to undue wear and strengthening of certain quite secondary roads, and force a body of workers to live a life of comparative isolation—all merely for the selfish advantage of the employer. This sort of thing will not do. Limited grouping, at least, is necessary so as to provide a definite traffic centre so that passenger transport can be more economically provided for the workers, and so that goods services by rail or road can be efficiently developed, by the units served making profitable groups for delivery and collection arrangements. Just as ribbon development is undesirable for house property, so is sporadic development for industrial units. It is true that transport nowadays can get there, but it might serve its purpose far better if travelling less miles and working among groups of factories.

Regional Distribution of Commodities.

Reverting now to the problem of regional distribution of goods throughout the country, touched upon a few paragraphs back. The point at issue can be more deeply understood by reference to an amazingly interesting study by Mr. John B. Andrew of the distribution throughout England of the principal imported commodities (*Manchester Guardian Commercial*,

January 30th 1932). Mr. Andrew uses tireless patience to prove—and does prove—that there is enormous waste in industry through payment of unnecessary freight charges through lack of skill in deciding ports of entry and routing products thence to consumption points. Britain is better equipped with useful sea-ports than most other countries, but Mr. Andrew argues that we do not use them properly. Official returns show the volume of Britain's import trade. Mr. Andrew goes much further, and by showing how the main commodities were distributed between the ports he dramatically exposes this waste. First, by a very careful analysis of population and a study of internal transport facilities, he assesses the proper hinterlands of each port. Then he takes the percentage which the population of each hinterland bears to the total population of the country, and argues that the figure thus ascertained should be a guide to the percentage of each staple article which ought to enter England through the particular port. Thus the Mersey hinterland, with 27 per cent of the population, should do 27 per cent of the import trade of the country and of each constituent article. Similarly with the Humber ports (15 per cent), London and the south-eastern ports (41 per cent), south-western ports (13 per cent) and north-east coast ports (4 per cent).

To examine Mr. Andrew's statements is to realize how far we fall short of that rational ideal to-day. Much of the dairy produce, for example, comes up the Thames ; little goes to the northern ports. Most of the fish supply arrives at the east coast ports. London handles nearly all the tea and coffee and the greater part of the beef, mutton and pork. On the other hand, the Mersey ports handle more than their due proportion of cattle and sheep, sago, ham, lard, raw cocoa, bananas and molasses. Bristol alone handles 54 per cent of the total imports of bananas, but the south-west generally gets far less than its share of all other fruits, but far more than its share of refined sugar. The Humber and north-east coast ports suffer because of the undue shares which go to the Mersey and the Thames.

Wheat Distribution a Good Test.

By way of a check upon the basis of these calculations Mr. Andrew examines with especial care the distribution of a commodity which stands a good chance of being supplied to different areas according to their needs, viz. wheat. As he says, wheat provides easily the best example, for it is of great bulk, nearly all of it comes from abroad, and from countries so distant that the trade is not confined to some particular short sea route ; and most of the trade is in the hands of great milling firms at the ports of entry, who can be trusted to avoid wasteful over-carriage. Actually, making allowance for the re-export of grain and export of flour milled from imported grain, the country's imports of wheat in the last available year were shared thus :

Mersey ports .	.	.	.	.	27·8 per cent
Humber ports .	.	.	.	.	16·7 ,, ,,
South-eastern ports .	.	.	.	36·6 ,, ,,
South-western ports .	.	.	.	15·4 ,, ,,
North-eastern ports .	.	.	.	3·3 ,, ,,

When allowance is made for the chance of slight errors through overlapping hinterlands, these practical results strikingly confirm the basis reached by Mr. Andrew, whose detailed findings deserve careful study.

Of course it is realized that different ports have specialized in handling certain products, and they face different sources of supply. But since sea transport is far cheaper than land transport, per ton-mile at any rate, varying the port of entry more in accordance with ascertained population might give further big savings in transport costs and a big step forward towards really scientific distribution.

Important New Trunk Roads.

Clearly, when an era of planned comfort is upon us, very important fresh road projects of a trunk nature will be approved almost without demur and rapidly proceeded with. Such vital links as that planned between London and South Wales will save many miles of travel between, say, London and Cardiff, giving a transit in every way more efficient than that now available, and opening up valuable new areas to agriculture, industry and residence. It is tragic that there is to-day no good main road between London, the capital, and Southampton, one of the leading passenger ports, and the strong suspicion is that since the docks at that port are in railway hands, strong political influences have discouraged efforts to put this right.

Then again, under the conditions into which we are rapidly moving the scandal of the " weak bridges " will be made short work of. A big effort has recently been made to put this right, and the Ministry of Transport is prepared to grant from the Road Fund up to 75 per cent of the cost of bringing up to proper strength those bridges over railways and canals which are, and under existing law are likely to remain, in an inadequate state for current main road traffic.

95

But more pressure is necessary to get the county authorities to act in this matter with the necessary energy. The suspicion is that some county councils are deliberately leaving the matter alone, on the theory that a " weak bridge " definitely lowers the capacity of a road for through traffic, and that if the bulk of the heavy traffic which would wish to use a certain main route has to use an alternative route, the resultant burden of road costs will fall upon a neighbouring body of ratepayers, to their detriment but to the political advantage of the clever folks who have engineered the saving of the rates in this way. Railway opposition on this topic must be overcome, for the planner is just as much concerned to avoid obstruction to the railway traffic through a bridge disaster as with road traffic's interests, and the solution of the problem cannot be left to the settlement by ordinary standardized grants in easy cases, and a *non-possumus* attitude in the difficult cases.

Will Air Transport develop Quickly?

It is too early yet to speak with any definiteness of the future distribution of our population about the country except that, for the reasons outlined, it will decline to be concentrated in a few spots and leave the rest almost empty. Possibly many true satellite towns will spring up, as they are already doing, within twenty-five miles of the main centres of population, and some of the pressure will be removed from those centres. But the chief factor of uncertainty lies in the air possibilities. Slight further technical progress with the autogyro would permit achievement of the earlier visions of Jules Verne, George Griffith and the rest, and light 'planes kept in the backyard would

96

become almost as common as light cars to-day. Then a technique of air traffic control will have to be worked out, landing grounds in the middle of cities—on station roofs, perhaps—would become commonplace, and before long one might expect the bulk of long-distance passenger travel to be by air, and most short distance travel by coach or car—excepting only such mass suburban business transport as continued necessary. This prospect must be before the minds of railway directors ; but their plan, surely, is to give much more study to the better working of freight traffic so as to make it pay its way at a lower level of charges and leave the railways unruffled by fears of future losses of traffic to air transport.

Shall we have to face much less export trade ? How will this alter traffic flow ? There is no space in this chapter to do more than put the questions, but that they are real problems which no official body to-day can answer—but which *must* be answered if planning is to be intelligent—need not be stressed.

CHAPTER SIX

TRANSPORT FACTORS IN TOWN PLANNING

ALTHOUGH in the previous chapter an attempt has been made to indicate on broad lines the regional planning tendencies which ought, by all logical reasoning, to affect the disposition of British population in the future, it is idle to assume that events will necessarily move in the directions indicated, at least without some bolder and more fully informed efforts than are yet seen. Evolution does not always mean an uncontrolled but natural movement towards *better* things ; the converse is equally possible. All the facts of recent economic history *do* indicate—we have to face it—a growing tendency for people to cluster into cities and towns. Britain is not alone in this matter. In 1925, two-thirds of the people of Germany were found to be living under urban conditions, in 1926 half the French people, despite the strong peasant bias of the population. In 1928, two-thirds of the population of such a new country as Australia lived under urban conditions, mainly in the five capital cities of the Dominion. The U.S.A. in 1930 showed over one-half and Canada in 1931 the same proportion living city lives, while the 1931 census of Great Britain gave the high figure of 80 per cent.

It would seem that only those people tend to live on the land who are essential to its cultivation or exploitation, and, tragically enough, under our cranky system they get the narrowest margin of profit of any class for their labours although they are the foundation of our civilization. In 1898 Sir William Crookes forecast that by 1931 the world's wheat supplies would not suffice to meet the demand ; yet in that year the carry over from one harvest cycle to another almost equalled a whole year's needs ! In but thirty-three years the scientists' researches and engineers' work have expanded the area of the corn-lands of the world, doubled the output per acre and halved the labour required. In consequence the rural population steadily grew less and the urban aggregations grew vaster.

Large Concentrations of Population.

It is popularly believed that the motor-car has checked this tendency. In America's prairie states there is a car to every four or five people, and nobody thinks much of going 50 to 100 miles each way for pleasure or interest. But the motor-car has not stopped the denudation of the country ; it has merely spread out the city more thinly by urbanizing wide belts of country on the fringe of the city until the break between city and country is blurred. Thus a host of new problems are created in the planning and sustentation of cities and the organizing of their transport. Population continues to concentrate in large units, and the country to grow more deserted. And, unfortunately, recent census returns show that the large cities tend to flourish at the expense of the small cities. The small city seems unable to compete—its industry drifts away unless rooted in some local circumstance—its amenities prove less attractive or advantageous. Mass seems to attract mass, and there is some ground for arguing, with Mr. Frank Pick, that economic conditions appear to grow more favourable with size as, e.g., in railway rates and charges, or railway services and facilities. The small city is handicapped. But the extravagant growth of the five largest cities of Western civilization—Paris, London, Berlin, New York and Chicago—has only taken place in the past sixty or seventy years, coinciding with the period of most intensive transport development. In growing to their current immense dimensions they have provided the transport expert with truly terrible problems in the provisioning of such cities and the maintenance of their internal circulation so essential to health and productivity.

While, therefore, the regional planner is working hard for a reversal of this process, for the creation of satellite towns and a general decentralization of life, and while modern science increasingly makes such benefits possible, there is an immense momentum to be overcome both in the instincts and upbringing—often narrow and misguided—of our citizens themselves, and in the interplay of economic forces which, in the present chaos, work for a perpetuation of the present unholy muddle and even make it worse. So that the expert, while he may study, speculate and exhort his fellows as to what ought to be, in his own and cognate spheres of interest, must also be prepared to deal as positively and helpfully as possible with our present muddle. This chapter, therefore, while offering no shadow of defence for the mad state of affairs around us, begins by seeing them as they are and offers suggestions for the next few steps out of the tangle.

City Transport Constantly Increasing.

City transport falls into two categories, points out Mr. Pick. There is the transport which is needful to enable the city to come into existence, which brings from afar first the people and then their supplies, and takes away the goods they produce. And there is the local transport required for the movement of people from place to place within the city and its vicinity, and for the collection and distribution of the goods they want. Local passenger movement in cities, it is strange to note, grows more intense as the city grows more populous. While the regional city on an average travels only from 200 to 300 times each year per head, London, even in 1930, travelled 518 times and New York 575 times. Clearly, size stimulates movement. The rate at which life is lived is quickened.

In 1905, when the Royal Commission on London Traffic issued its final report, local passenger journeys taken in the Greater London area averaged 200 per head. By 1914 the number had risen to 311, and in 1919 to 413. In 1924 it was 440, and in 1929 511, and in 1930, as stated, 518. The later growth came as the transport facilities increased. While there was during the period constant complaint of local transport shortage, nobody could have guessed that so huge a movement was suppressed as was found there when facilities were expanded. Between 1914 and 1930 the total number of local passengers carried in Greater London by all forms of conveyance increased by 1,812 millions, or 78 per cent, while the population had only increased by 7 per cent.

This feature causes providers of local city passenger transport to despair. New facilities are constructed to relieve congestion, but in fact they soon lead to new and ever greater congestion. Between 1921 and 1931 Greater London's population increased by 723,000 people or nearly 10 per cent, but this was a balance figure of many increases and decreases, and hid the true redistribution of population. Again, population followed rapid transit facilities. The underground extension to Edgware added 87,000 or 128 per cent to the population of the district, or 12 times the average rate. The electrification of the Southern Railway combined with the L.C.C. Housing schemes has largely swollen the population of South-East London.

Migration from other parts into South-East England during the decade has been considerable. Of 615,000 people who arrived, 210,000 settled in Greater London, and probably London was, in fact, the attraction for many more. So that the transport problems of London in particular, and all important cities next, specially call for greater study than they yet receive, based upon better factual knowledge than is to-day to be had and taking cognisance of vast plans for an improved city in the future. The manful efforts of those who are now trying to bring order out of chaos deserve at least to be understood. A better-informed electorate can help them, for we do not agree with a despairing remark of Mr. Pick's, " The Metropolitan city is the key problem in civilization and it still escapes a rational solution."

More Factories for London.

Dr. Raymond Unwin shows that between the end of the War and 1930, 1,145 entirely new factories have been built within the London region, employing 131,569 people. These people, with their dependants, mean an extra population of over half a million.

London has added to itself a fresh industrial city which, if distributed into five towns of 100,000 persons each, would have presented a more manageable mass of people and certainly one which might have been handled to yield a better measure of amenity.

About a third of the total passenger movement in Greater London is of workers moving between home and work due to the wide dispersion between housing and industry. Surely an unsatisfactory feature, arising from bad adjustment of housing and industry which, in a planned metropolis, might largely be avoided. Of course, instances going against this thesis can be noted. Acton turns out daily 13,500 workers and takes in 14,500 from elsewhere. Hammersmith turns out 32,000 workers and takes in 29,000.

Since 1920, there have been immense but quite unco-ordinated housing developments in Greater London. Though the population has re-sorted itself the rides taken per head have gone up rather than down. Partly this is because before the War one-third of the working-class families moved each year, but since, due partly to rent restriction laws, the rate of removal has fallen to once in ten years. Whether the restoration of fluidity to the population is desirable is a matter of opinion, but there is surely no need to place industry anywhere but where the needed conveniences, transport, water supply, etc., exist or can best be furnished. The centre of a city or a residential quarter need not be spoiled by the intrusion of industrial concerns.

City Transport must be Improved.

Yet in all metropolitan cities there is an insistent demand for more facilities. Those which exist tend steadily to become congested, and the same fate over-takes those built in relief. The burden of keeping pace with demand is difficult, and it is never satisfied or relaxed. Clearly any solution of the question of growth in cities must mean bulking together the problem of transport with all other problems of city planning and management. For centralization, we must repeat, is not only not necessary, but tends now to be definitely uneconomic—at least for passenger transport. The industrial demand upon transport facilities is the most expensive demand. It represents the peaks of the traffic of the day for which rolling stock and equipment, staff and power, must be supplied for a short interval of useful service only. Transport development must be linked to housing development, industrial development, and other phases of town planning so that the acute burdens are met not by the multiplication of transport facilities but by the co-ordination of factory and house construction so that the worker can enjoy amenities at home yet not have to proceed so far afield for his daily work. We are surely up against the irony of " progress ", when the bare effort of living in a city exhausts the strength.

There are at length signs of a real wish to grapple with the problem in London. As a small portent, the " Green Belt " for London, being talked of five years ago at the Town Planning Institute, is now gripping the popular imagination—though it is a bit late in some directions to prevent most unsightly development of an almost vandalistic type.

Will Cities keep Growing ?

Must cities go on growing ? Are there any natural checks on growth ? Transport experts see two. One is that while the circulatory system of roads and railways may expand and multiply the heart of the city becomes its weakest point, and congestion at the centre slows down the whole pace of movement and leads to costly waste of time and uneconomic use of the transport plant. The other check to growth turns upon the average speed of movement. Although the supreme value of speed is realized at large in the world, it is the time limit set to the length of the daily journey for most people that may really govern the city's growth. Suppose 30 minutes travelling time with a short walk at either end to be the ·average limit. The speed of local railways has in London crawled up from 14 to 25 m.p.h. and through express trains with limited stops the city is already stretching to the 25- or 30-mile zone, at any rate with scattered groups of housing about the stations. The express coach came to widen the housing spread and, for lack of courageous handling of finance, to make ribbon development attractive to the builder. Under special conditions thousands move daily from Southend and Brighton and neighbourhood to London, and from Blackpool and Southport to Manchester, though most of these get more than the average reward for less than the average length of daily labour.

London's Peculiar Problems.

London's planning and transport problems may be further studied as being of primary importance to the nation and typical, to a degree, of conditions elsewhere. There are enthusiasts for overhead roads, trolley buses, goods tubes, road widening, central freight terminals, and many other specifics, but it seems certain that no single panacea will solve all

London's problems. There must be vigorous yet adroit application of many different improvements each in its proper place, and some phases of the situation may always require very special treatment. The basis of the whole structure, however, must surely be a plan which allows for roads adequate to the city's present and future needs—which simple truism plunges us headlong into another set of problems and confusions.

All industrial towns have suffered intensely for the lack of Town Planning powers, the first Town Planning Act of 1909 being too late by eighty years to safeguard their rapid development period. Even when the principle of controlling development for the common good was conferred by that Act, it was confined to action in regard to unbuilt-upon land likely to be developed. Although the unfortunate results of that delay were almost immediately obvious, it took a further twenty-three years to get control extended to built areas—twenty-three years during which very much has been done actually bad or potentially prejudicial, such as the perpetuation of narrow and tortuous streets ; want of proper and direct road communications ; bad layouts of, and to, many road junctions ; absence of open spaces ; incongruous association of buildings both in regard to usage and design ; and in many cases the extension of such unsatisfactory development ; all of which has tended to make the work of the Town Planner harder and much more costly now that at last such extended powers of control have been obtained.

Make Roads to fit the Traffic.

However, under the Town and Country Planning Act, 1932, some measure of control may at length be exercised over built-up areas, and the preparation of a planning scheme for Greater London becomes legally possible, although the powers sought in the Bill were somewhat whittled down in the Act. Major H. A. Crawfurd, a Patron of the Industrial Transport Association and a member of the Royal Commission on Transport, 1930, has made a special study of London transport problems, and his main conclusion cannot too often be emphasized. It is that we must fit our road system for the traffic it is called on to carry, instead of being content to go on trying to compress ever-growing traffics to our limited streets. The London and Home Counties Traffic Advisory Committee, set up in 1924, should have tackled this matter at once but failed to take a single practical step in that direction. Four successive Ministers of Transport between them contributed exactly nothing to the solution of the problem. The Ministry of Transport, absorbed in the task of devising means of restricting transport, had no time to take steps to facilitate it. In London's main streets, during business hours, traffic delays are scandalous and costly, for what lies at the root of the problem has never been tackled.

The Roman roads were built 1,600 or 1,700 years ago to meet adequately their needs. Their main lines still exist, and since that time there has been no first-class administrative measure calculated to fit London's roadway system as a whole to the traffic it is called upon to bear—until the appointment in 1935 of Colonel Bressey to investigate the problem.

Old but Sensible Laws.

An Act of 1667, passed after an agitation following the Great Fire of London, contains much that might have been remembered when, at the end of the War, men's minds turned to the general problems of reconstruction. The preamble spoke of the narrowness and incommodiousness of certain streets, and the need to enlarge certain named streets in the neighbourhood of St. Paul's Churchyard, and Section VI laid down, in simple language, the vital principle of " betterment ". A payment from those owners of property who would gain increased rental values was provided for, and was to be used to recompense such owners of houses and lands as were deprived of part or all of their property because it was converted into " streets, passages, markets and other public places ".

Major Crawfurd has unearthed another Act (of 1759) dealing with the development of London, the preamble of which states

> Whereas certain streets . . . within the city of London . . . are too narrow for the passing and re-passing as well of foot passengers as of coaches, carts and other carriages to the prejudice and inconvenience of the owners and inhabitants in houses in and near the same, and to the great hindrance of business, trade and commerce. . . .

So although in the days of George II, as in those of Charles II, Parliament recognized that the city of London was inadequate to the traffic even of those times, to-day, with a Minister and Ministry established expressly to promote transport facilities, the authorities are only beginning tardily to realize the imperative need for measures of the same sort.

Traffic Needs of Towns Neglected.

The motor vehicle, as a common means of conveyance, dates from about the beginning of the century.

108

Its use developed fast, and brought a revolution in travel comparable to that due to the invention of the steam engine a century earlier. In 1909 Parliament recognized this and took measures to cope with the new demands on the road system. These efforts were multiplied after the War, and there has been much rebuilding of old roads in rural areas and a wholesale building of new roads. But the task of dealing with the traffic needs of towns has been neglected, especially in the case of London. Little has been done and even less planned. So to-day London is the centre of a vast system of new and improved trunk roads, which stop short at its edge and pour a vastly increased volume of traffic into its narrow and congested streets, where it cannot possibly circulate freely. It is as though a bath were fitted with a dozen new taps, and no extra provision made for the water to flow away. When the taps are turned on, inevitably the water overflows. Typical are the cases of the new Great West Road, Chiswick High Road and Hammersmith Broadway, and the series of by-passes which converge on to Finchley Road, Swiss Cottage and Baker Street. All round London this is repeated.

Equally many instances could be given of places where the convergence of several main routes at one spot calls for a wide area where the traffic streams meet—but instead the streets narrow still more ! Think of Hammersmith Broadway, Elephant and Castle, Camden Town, Angel Islington, Shoreditch, Dalston, New Cross, Oxford Street, Tottenham Court Road, and so on.

Still worse ; during the past fifteen years, while the traffic volume has so increased, there has been much rebuilding in some important London streets, but chance after chance has been neglected of conforming to a scheme that would have given needed space. Instead, new buildings have been erected on exactly the sites of the old, and new and greater obstacles set up against future improvement.

Piecemeal, Unrelated Improvements.

There have been piecemeal improvements, unrelated to any general scheme, so that improvement in one place has often meant added congestion in another. Much of the talk about a new Charing Cross Bridge is very short-sighted. People speak as though the building of a bridge at that point would solve some traffic problem. It might, indeed, relieve the weight of traffic on bridges to each side of it, but if the extra vehicles it carried were decanted somewhere about the Edith Cavell memorial, congestion in Charing Cross Road would be much worse—unless this or St. Martin's Lane were widened very much. To build the bridge without dealing with the areas it would serve would be foolish. This, though, is what we usually do. Years ago King William Street was constructed— and added to the vortex of traffic at the Bank. Later, Queen Victoria Street was cut through and congestion at the Bank grew worse. In recent years the Strand has been widened—and now it can pour four times as much traffic into the confined channel of Fleet Street. Even Kingsway, the one major street improvement in London in our time, leads to a bottle neck at the Holborn end. This is characteristic of the London street system ; a series of constricted points makes for the continual hold-up of traffic, and, instead of a steady stream of vehicles, one sees from above a succession of alternate lengths of streets comparatively free from traffic, and, between them, areas hopelessly choked.

While London has grown anyhow, Paris has been boldly planned as a city intersected by broad boulevards in every direction from the outskirts to the very centre. So, while in London along many chief streets for many hours of the day you can but crawl, in Paris movement is free and delay in vehicular traffic a rarity. The simple difference is that for nearly seventy years Paris has been methodically developed according to one general plan. A major improvement in the very heart of the city completed six or seven years ago accords strictly with the plans elaborated by Haussmann and his associates in 1868.

Bold Planning Needed for London.

This is London's supreme need, and Major Crawfurd maintains that the present Minister of Transport, by taking the preliminary step of arranging for a survey of traffic needs, has done one thing worth more than all the legislation that has emanated from the Ministry since its formation.

But without other and bolder steps this beginning may be in vain. The research must lead to the completion of a detailed scheme of improvements covering the probable needs of the public for road traffic facilities for the next fifty years, including the widening and improvement of existing roads, the provision of more space at main-road junctions, the cutting of new traffic avenues, intelligent traffic regulations and pedestrian safety.

One authority is needed, with over-riding powers to put the scheme into operation. Such powers as might be used to obstruct or hold up the scheme, at present residing in other bodies, municipal or otherwise, should be transferred to the chosen authority. Some say the proper authority should be a small committee of the L.C.C., plus one or two co-opted or nominated persons. The present L.C.C. (1935) seem to have solved this part of the problem to their own satisfaction, though how far they can go with the bold schemes they are now putting forth, without encountering dangerous opposition, remains to be seen. One wishes them every success, however.

It is often urged against any contemplated scheme on this big scale for London, that it would be too costly. There are two answers, both complete. The first is that it will have to be done some day, and anyhow the sooner it is begun the less will it cost. The second answer is, that it is very doubtful if it really need cost anything at all. Looking ahead for fifty years, plans can be laid so that much of present-day through traffic can proceed along alternative routes cut through neighbourhoods where property values are lower. This would relieve the present roads and create values elsewhere by the improvements made. Think, for instance, of a road cut, largely using existing thoroughfares, north of and parallel to Oxford Street. As London's business centres are expanding in many directions, especially northwards and north-west, the creation of new streets will prove so attractive that greater values will be created. Thus, enormous accretions are possible to land values, which can all be used for the purpose of the next step in the development of the land scheme. Suppose a capital sum was needed to start the scheme. £50,000,000 would be a small payment, even if made outright, for the benefits that would follow. But the sum needed by way of interest and sinking fund for the repayment of this amount can easily be found if the annual theft from the Road Fund is stopped.

Central Authority for London's Roads.

New administrative problems would arise as these improvements were carried out, and decisions would have to be taken as to their upkeep and maintenance when constructed. For several reasons, one being the need for uniformity of surface, the main streets in the whole London roadway system ought to be maintained and controlled by a central authority, the London County Council.

Slow, inconvenient or costly transport injures the commercial and industrial life of the country, and indirectly the whole community loses. Despite certain improvements, the congestion along the whole length of the London dock and wharf area still causes immense waste. The endless delays to commercial vehicles in the central streets means an annual direct loss in time and fuel of tens of millions of pounds. Mr. W. H. Gaunt, O.B.E., M.Inst.T. (Messrs. J. Lyons & Co.'s Distribution Manager), told the Royal Commission on Transport that he had based figures on the difference between the time taken by delivery vans to make a call, and the consequent cost per hour or per call of the van, in the hours when the streets are comparatively free and when great congestion is found. In the former case his average cost per call was 3s. 4d. and in the latter, 6s. 8d. Probably, too, this means the employment of very many more vans by this concern and others similarly circumstanced, to achieve all their needed deliveries, and so fresh vehicles have to be bought and put on the streets to add to the congestion.

Eliminate the Horse ?

The rigid compression necessary throughout this work was nowhere more of a burden than in drafting this chapter ; so many tempting ideas and projects crowd upon one and deserve brief mention and an attempt at fair assessment. The favourite plan with many people, the entire elimination of the horse from busy city streets, has much to recommend it now that " mechanical horses " have overcome their early technical difficulties and appear sound and reliable jobs. Overhead roads based on American practice have some exponents, but one does not see any case in which that practice need be imitated in this country if the re-planning advocated earlier can be undertaken. A London goods tube railway, on the other hand, gaining much of its inspiration from the Chicago traffic tunnels, has very much to recommend it if only its promoters can be brought into co-operative association with the main-line railways to such a useful end that the electrification of at least the urban portions of the main lines, plus a joining together underground of the present dead-ends, plus some linking with the goods tube which will make, if not for some through running, at least for a speedy and cheap transfer of freight, will be achieved.

No Steam Trains in London !

Whether or not the goods tube railway matures—and it is a most intriguing scheme having a sharp appeal to one who has journeyed through the Chicago goods tunnels—the point made about electrification of main-line railways, at the very least so far as they are within the urban areas, stands as essential to a decent London. All passenger trains should complete the last twenty miles or so of their journey into the Metropolis under the charge of an electric or diesel electric locomotive. Goods trains that now penetrate into the heart of London should also be electrified in order to pass through the connecting tunnels between the different termini. Railways in cuttings could then be roofed over for a large part of their length, as Mr. A. C. Bossom has so consistently pointed out, and building space of immense value would be thus created. Thus not only would there be a very inexpensive way of hastening slum clearance by providing good and conveniently-placed decanting sites for the comfortable housing of people who desired or needed to remain near their work in London's centre ; but when the aeroplane is ready to rise and descend in a more nearly vertical plane—and progress with the autogyro is already remarkable—some flat-topped sites near the real destinations of air passengers can soon be available.

Any comprehensive plan for London envisages a definition of London within a certain boundary, a green belt which shall for ever provide lungs and recreation space for us. On this green belt must be created aerodromes, and within the city we must provide, for want of a better name, cab ranks for air taxis. These newly-covered sections of main railway will provide some, at least, of the taxi ranks, and there will be no lack of enterprising storekeepers who will provide others.

Mr. A. C. Bossom's Valuable Plan.

But the important use of these new sites as " decanting areas " must be further stressed. Mr. Bossom, from first-hand experience as Chairman of an L.C.C.

sub-Committee handling the acquisition of property known as unhealthy areas, points out that the earlier clearances and improvements achieved by some borough councils and the L.C.C. make it practically impossible now, except at great delay, to acquire new property anywhere in these slum-congested districts, so that people are just shifted around on a very slow-motion basis. Hence Mr. Bossom has become the chief exponent of the plan of using many of the 350 acres of sunken land now only covered by railway tracks. On the other hand, some 160 acres of land exist on which are housed at least 60,000 people in highly unsatisfactory conditions. Experts have combed London, and land for the decanting operation is not found. Yet here are these 350 acres of land doing nothing but allowing trains to run about below ground level and, from a height of 16 feet to the sky, no use is being made of it. There are about 40 acres in the general Paddington area, there is a lot more in Hackney, Farringdon Road, around York Road, Kensington, and about the docks. Railway directors bemoan hard times, while this gold mine is ignored instead of worked in co-operation with the municipal authorities.

The suggestion is that over these open tracks a steel and concrete mat be constructed, 16 feet above them, and, upon this, thoroughfares to lead from the heart of the city to its perimeter, and housing work ; indeed, many sorts of buildings might be constructed to supply definite needs and remove at once the restricting influence from the slum clearance problem. Conveyance upwards of noise from the trains would be overcome by a double air space in the concrete mat. Vibration would be eliminated by introducing under the base of the steel columns where they stand on their concrete foundations, a three-quarter-inch mat made of sheets of lead, asbestos and a piece of steel. This plan has been followed continuously and most successfully in New York, Chicago and other American cities, and with the necessary adjustments from the known costs there it is computed that from £6,000 to £10,000 per acre would be the cost of creating this new land right in the heart of the city. By this plan, the grave delays to be faced in ordinary slum clearance programmes are eliminated at the source, and the price paid would compare well with £30,000 an acre and more which has been demanded for land for re-housing slum-dwellers.

Using London's Rivers and Valleys.

Planners in London do not forget the great influence of the Thames upon the original site and the development of the city, and they want to continue taking advantage of the river's shape and usefulness. The Thames Embankment is the best way out to the south-west, but it should be extended westward, with a new bridge to the Portsmouth Road at Wandsworth. The South Side Embankment must be completed, as the L.C.C. well realize, and there is room for the reclamation of hundreds of acres of mudbanks. The centre of London is not the place for a tidal reservoir, and there is no reason why the river at Waterloo should be a hundred yards wider than it is at London Bridge.

The Brent Valley has already been most useful as a route for the North Circular Road, and the Beverley Brook provides the route for a new parkway to Richmond Park. The Wandle Valley is capable of much the same treatment, and the Lea Valley is still open to provide a magnificent parkway, direct from the heart of the congested areas each side of it, right out into the open country and Epping Forest. The water reservoirs which it contains would be of enormous scenic value in such a parkway. Valuable industrial sites would be opened out by the creation of such a boulevard, linking up with the new Victoria Docks approach, but to complete the scheme it should be linked up across the river by a Woolwich bridge to the South Circular road. The Roding Valley asks to be used as a link between the Barking Bypass and the North Circular and Romford roads.

The Pool of London, famous for shipping until it moved down the river, is still available for fast motor-boats, and its use as a landing place for flying boats is certain to be extended in the near future. Its possible use for this purpose must affect the number and position of new bridges which must probably be allowed for soon.

There is much to be said for the crossing of main streams of traffic, wherever practicable, at different levels. This is only possible by the use of the river valleys and special points where levels permit. The bridging of the Strand at Wellington Street is an instance.

Railways are roundly condemned by some city planners as the greatest obstacle to a satisfactory plan of surface development, and it is argued that they must inevitably be put underground throughout the whole central area of London. This would certainly set free a number of potential routes out of London, e.g. Ludgate to Dulwich, Fenchurch Street to the East, London Bridge to New Cross. Supposing, too, the railway from Charing Cross to Cannon Street and

London Bridge could be removed, the way would be clear for a new southern boulevard of immense value in distributing traffic south of the Thames.

Why Stick to London?

Many of the improvements suggested, however, are admittedly but palliatives. Why should not an alternative to London be planned? Even if it were desirable for the administration to remain in London, many industries might be drawn away and placed in more suitable locations, and to plan for siting these industries and so creating new and well-considered satellite towns does seem by far the simpler as well as the more economical and really health-giving and comfort-bringing solution for the great mass of citizens, whose happiness in life should be the main aim of all reformers. We must choose quickly between our sentiment for the cattle-tracks which have become streets and our determination not to wear out our lives, through noise, congestion and discomfort, having achieved only a small part of what is in us to do.

CHAPTER SEVEN

ROADS TO MEET MODERN NEEDS

THE average man takes the country's roads for granted and forgets their vital part in maintaining our comfort and, indeed, in making possible our life as we now live it. After the Romans left Britain (about A.D. 411) the roads, like the rest of the country, fell into a bad way. Liability for road construction lay with the local landlord. Through traffic hardly existed, and the peril and discomfort of travelling were such that people were content to live and die in the vicinity in which they were born. Travelling was practically restricted to the nobility and clergy. It was the practice to lay in provisions on the approach of winter, as though preparing for a siege. Centuries passed without material change, and only in 1555 was a general Highway Act passed requiring inhabitants of a parish to perform what was termed "statute duty" or compulsory labour to keep roads in repair. Highway surveyors were appointed, but the office carried no salary, and most of those so serving knew little of road construction. Engineering difficulties were avoided by taking the line of least resistance, and so there was little or no systematic development of British roads until the passing of the first Turnpike Act in 1663. The trunk roads then existing became known as Turnpike Roads, under control of Turnpike Trusts empowered to erect toll gates and take tolls from passengers for road upkeep. Yet at Common Law the general inhabitants of the parish were still liable to maintain the highways in the parish, and "Statute Duty" continued until abolished by the Highways Act of 1835, which charged the cost of highway administration to the parish rates.

The turnpike system failed, despite the heavy tolls levied, in its later years doubtless because of railway inauguration which meant that practically all the national traffic left the roads. The 1835 Act also failed, and though all admitted the unsatisfactory state of the roads and their control, and bills were produced year after year, nothing material was achieved until by the Highway Act of 1862 highway districts were formed, each comprising a number of highway parishes. The powers of the old highway surveyors in those parishes passed to new Highway Boards who could appoint and pay officials. Since the Local Government Act, 1894, the country has been divided into Urban and Rural Districts administered by Urban and Rural District Councils.

Divided Control of Highways.

Roads dis-turnpiked after 31st December 1870, and any other roads specially so declared became known as "main roads" and in general were maintained by the County Council. Urban authorities, however, had a limited option of retaining control of main roads in their district and receiving an annual payment from the County rate. The County Councils may also, by agreement or otherwise, delegate the control of main roads to the highway authority of any highway area, making an annual payment from the County Rate towards the cost of the undertaking.

The Authorities exercising highway powers in England are, then :—

In respect of main roads :
  County Councils,
  County Borough Councils,
  Town Councils and Urban District Councils that have " claimed to retain " under the Local Government Act, 1888, Sec. II (2), and District Councils that have entered into contracts with County Councils under the Local Government Act, 1888, Sec. II (4).

In respect of other roads :
  County Borough Councils,
  Town Councils,
  Urban District Councils and Rural District Councils.

Gradual development of motor-vehicles forced highway authorities to take road maintenance and construction seriously, and in 1908 tar was first used in connection with road maintenance and the era of the old waterbound methods was practically ended. About this time the general use of tar both for spraying and for tar-macadam was begun.

Creation of the Road Board.

Changes in the traffic and in methods of road construction forced attention to the inadequacy of highway administration and, by the Development and Roads Improvement Fund Act, 1909, the Road Board was constituted. Duties were levied on motor spirit, and the sums so raised, together with the proceeds of duties on motor-cars and carriage licences, were administered by the Road Board by way of grants to the highway authorities : though it will not be seriously claimed that the grants were at all adequate to the real needs.

When the Ministry of Transport was formed its Roads Department took over the functions of the old Road Board. The Ministry also administered the Road Fund into which, under the Roads Act, 1920, was paid the proceeds of the tax on all mechanically propelled vehicles using the highways. Light cars paid on a horse-power basis, and heavy motors on seating capacity or weight. The Ministry further undertook a classification of the country's roads into First, Second, and, later, Third Class, according as they were deemed to be (1) main roads, (2) district roads or borough roads, (3) selected unclassified roads, chiefly in rural areas. Grants were made for ordinary road maintenance or reconstruction works of 50 per cent of approved outlay on First-Class roads, 25 per cent on Second-Class roads, and 20 per cent for " Selected Unclassified Roads ". For special improvements, trunk roads, and big bridge schemes, the Ministry made grants varying in percentage of approved outlay up to 100 per cent.

Highway authorities viewed with alarm, however, the increase of local rates to meet increasing demands for highways, and agitated for more aid from the Road Fund. From a different angle the British Railways, with skilled publicists and well-organized political supporters, were agitating for radical changes in road and road transport financial arrangements, and the Salter Committee which sat in the summer of 1932 showed much sympathy with both types of complainant. By 1932, they said, there were in

Britain about a million private cars, 627,000 motor-cycles, 364,000 goods motor vehicles and 87,000 taxi-cabs, motor omnibuses and coaches, and the internal combustion engine had placed a mechanized horse at the service of each man and woman for the carriage of themselves and their goods. In no other country was the number of cars so great in relation to area.

Statesmanship.

In the House of Commons on April 24th 1928, by the way, Mr. Winston Churchill, then Chancellor of the Exchequer, had said, "It is the duty of the State to hold the balance even between road and rail." Whereupon he "raided" the Road Fund (money contributed by road users under agreed conditions) for the benefit of the general revenues of the country !

To return : the Salter Committee ventured, on grounds capable of grave criticism, to forecast the average annual cost to the nation of road maintenance and development at £60,000,000, and they debited £23,500,000 of this to commercial goods vehicles and £36,500,000 to all other mechanically propelled vehicles. They recommended detailed scales of licence duties on commercial vehicles designed to give the total figure laid down, and involving serious increases on vehicles which, as they said, enjoyed a "large concealed subsidy" through using a fuel or motive power other than petrol.

The essence of the bitter controversy which arose on the appearance of the Salter Report was that road transport did pay, in all, about £60,000,000 annually to the Exchequer, but only £22,000,000 or so of it was allocated to the roads. Local ratepayers had therefore to find about £38,000,000 annually for road purposes, while the Chancellor was applying to general revenue an enormous sum paid annually by road transport users, in defiance alike of all good faith and of any canon of sound transport economics. What a typical example of how the general mass of us will calmly allow a gross wrong and impudent robbery if carried through on a big enough scale and with a cultured air. By 1935 the Chancellor, about to make another raid, and secure in his big majority of Parliamentary votes, hardly felt any apology necessary but contented himself with mentioning that the Fund was rich—*of course it was*, for needed expenditures had not been permitted—and had been raided before !

Salter's opponents drew attention to the highly inconvenient piece of advice on road cost allocation given in the Final Report of the Royal Commission on Transport (issued in 1931), page 70, para. 249 : " We recommend that . . . in future one-third of the cost of the highways should fall on the ratepayer and that two-thirds should be borne by the motorist." Where the Salter Committee's recommendations differed materially from those of the Royal Commission it may be stressed that the latter was an unbiassed and numerically strong body, which gave many months of study to an enormous amount of evidence, whereas " Salter " heard no evidence and consisted entirely of directly interested parties.

Paying for Better Roads.

In 1900, i.e. before the introduction of the motor-car, the annual outlay for upkeep and new road construction was about £15,300,000 or, on the basis of 1925 prices, £23,700,000. The increase in road costs attributable to the advent of motor transport is therefore £45,000,000 or, on the 1925 price basis, only £36,000,000, and pretty obviously the community has at its disposal a greater mileage of better roads for various non-taxable uses than it had in 1900.

Broadly, the road interests' case in 1932 was that they should only pay the difference between the old and the new outlay, for persons other than motorists used the roads before motor traffic developed. And, since the revenue from the fuel tax (£31,000,000) and from the taxation of motor vehicles (£28,000,000) totalled £59,000,000, they held that motor transport already actually paid more than its share of the cost of road upkeep. The railway attitude was that motor transport ought also to contribute to the amortization of the initial cost of the roads, as the railways did in regard to their permanent way.

The impartial mind sees two flaws in the latter argument, each capable of much fuller development. First, the railways in fact do not amortize their permanent way costs, but go on trying to pay interest on the capital outlay thereon for all time. Second, that the road users' contribution figures could quite easily be held to prove that they at least subscribe a sum which would more than take care of any amortization of road costs, if it were properly applied to that end ; and that, anyhow, since roads are not exclusively owned by the road vehicle user, but by the community, it should not be his function to amortize road costs.

By the Road Traffic Act, 1930, maximum weights, dimensions and speeds of vehicles were defined, drivers' hours were regulated, and Area Traffic Commissions were created and empowered to inspect vehicles and to license certain kinds of passenger road vehicles. Provisions for the public safety, others regarding physical fitness and age of drivers, and compulsory insurance against third party risks were also embodied, but the 1930 Act did not provide for the co-ordination of road and rail services, nor did it stipulate the fixation of road transport tariffs. Despite heavy taxation, then, goods vehicle operators continued to enjoy some liberty, though much stricter regulation was imposed upon road passenger transport.

Salter Recommendations Implemented.

Though the Transport Minister invited interested bodies to comment on the Salter Report made public in the autumn of 1932 and they did so—very strongly— the taxation of heavy vehicles was materially raised in the 1933 Budget, and preparation of a new Bill based mainly on the Salter recommendations went quietly forward. Despite earnest opposition from trading interests, it became law late in 1933 as the Road and Rail Traffic Act, 1933. Its first part extended to goods road transport, the system of oper- ating under a licence issued by the Chairman of the Traffic Commissioners in one of the areas created in 1930. There are three types of licences, bearing the distinguishing letters A, B or C. The first or Public Carriers' Licence is valid for two years, and must be obtained by carriers transporting goods for third parties ; the second or Limited Carriers' Licence is only valid for one year and applies to vehicles carry- ing goods either in connection with the owner's busi- ness or for hire or reward ; the third, which is valid for three years, applies to manufacturers or enter- prises using their own vehicles for their own purposes —termed by Salter as " ancillary users ". To settle disputes respecting the issue of licences, an Appeal Tribunal of three members has been created.

Under Part III of the 1933 Act there has been created a Transport Advisory Council, to advise the Minister on questions of transport collaboration, etc. Part II, evidently added somewhat hurriedly to the main draft of the measure, empowers the railways to grant " agreed charges " subject to certain safeguards.

The 1933 Budget changes were foolish in that they have merely encouraged the use of more and smaller vehicles, which add to road congestion. The speed restriction of 20 m.p.h. for lorries over 2½ tons unladen, while lighter vehicles may legally do 30 m.p.h., works in the same direction.

Government Road Policy Railway Inspired.

Road interests say openly that the entire road construction and usage policy of the British Government is under the dictatorship of railway interests, through the four railway general managers working both direct, and via the Railway Companies' Association. Firstly, dealing with road finance, here are approximate figures for the financial year 1933–4 :

YIELD FROM VEHICLES :
Taxation and driving licences . . .	£30,700,000
Petrol duty	36,500,000
	£67,200,000

ROAD EXPENDITURE :
Maintenance, repairs and improvements .	£38,606,000
New construction, cleansing and administration	12,153,000
	£50,759,000
EXCESS OVER EXPENDITURE	£16,559,000

Of the £67,200,000 received in motor taxation, only

£25,500,000 was paid into the Road Fund, out of which £13,000,000 was paid in direct grants to road costs, £6,500,000 to local authorities, and £1,500,000 in administration and other expenses—£21,000,000 in all. Thus, £45,000,000 was retained by the Government for general purposes.

But, as the total road expenditure for the year was £50,759,000, the balance of £29,759,000 had to be provided by local ratepayers. The actual amount put up by local ratepayers may exceed that sum, but, taking it as correct, then the Government and Local Government authorities extracted from the country, by way of taxation in one form or another, nominally for road costs, £97,000,000 for the year under review, and spent only £51,000,000, leaving the vast balance of £46,000,000 used for other than road purposes. In short, for every £1 collected as road taxation, 10s. is used for purposes other than those for which it is intended. If that is not " double crossing " the public in the matter of road taxation, what is ?

In consequence, certain road interests frankly say that though the Ministry of Transport has, in theory, much and many powers, the Minister is but a figure-head. So long as he uses his powers restrictively against road transport, he is left to enjoy the pleasure of press publicity for his little " stunts ". More than once in public, Cabinet Ministers have said that railways must be saved. So a Minister of Transport who wants to keep popular with his colleagues doesn't stir up too much mud in the circles that count. Most efforts of recent Ministers have been to limit and restrict the use of the roads for their public purpose, either by direct control or by the indirect means of imposing heavy burdens upon the road user.

Bring our Roads Up to Date.

Surely but a modest new outlay would now make all existing first-class roads suitable for modern traffic needs, and that outlay should be made without delay. Our Class One roads, one submits, could easily be made quite capable of carrying almost any weight if distributed over a suitable number of axles, the wheels of which are pneumatic tyred. Nobody yet needs to see the whole of our 178,000 miles converted into highways for fast motors, but we do at least want and must have the 26,000 miles of " First-Class " highways made and kept really safe and suitable for modern traffic. Also, of course, the densely populated area routes need drastic modification, more so than is generally supposed. Given these, the minor routes would need correspondingly minor modification, whilst the picturesque lanes should be preserved as such.

But at Westminster we have a railway-minded Government. The railways already are in such a position that they can, and do, largely control road development and use in Britain, and through their splendid Parliamentary organization and their very clever, but pernicious, propaganda they seem to be rapidly re-acquiring complete monopoly. Westminster's tendency to " save the railways at any price " would seem to be conceived not so much in the interests of the public as of finance ; high or otherwise, as one prefers.

More Five-Year Plans.

The public should not be deceived by the showmanship of the Transport Ministers, as instanced by the announcement in January 1935, at the Annual Banquet of the Birmingham Jewellers' and Silver-smiths' Association, of a Five-Year Plan for Roads. Mr. Lloyd George was just capturing public interest for his " New Deal " plans and perhaps a counter-blast was needed by the Government. Anyhow, the Five-Year Plan for Roads proved, on study, to con-tain nothing that was not understood to be already in hand as required. The plan was to provide for improvements which highway authorities, thinking and arranging ahead, could reasonably be expected to carry out in the period. The Government would try to eliminate within the five years " all those weak bridges in the possession of railways and other statutory owners which most seriously limit the free flow of traffic ". (This is a fairly definite statement, but it concerns a glaring evil which has been persistently pointed out for years and so is a very tardy expression of business-like intentions.) The Minister also laid stress on the official purpose to provide, where needed, dual carriageways, footpaths and cycling tracks, to remove blind corners, to circumvent the dangers of cross-roads and to reduce the camber and improve the super-elevation of roads, grants being contingent upon the carrying out of such works " on an adequate· and modern scale ". He refuted the idea that there was wide scope for new road construction in this country, saying that most highway work financed with the aid of the Road Fund must be in the recon-struction or adaptation of old roads to new traffic conditions.

The author dissents flatly from that last statement, seeing in it clear proof of the assertion already made that the Ministry, as now organized, is chiefly one to

131

safeguard *rail* transport. The phantom crisis of 1930 gave a splendid excuse for practically stopping road improvement work, and for four years very few road and bridge improvement schemes were initiated or financed. Traffic has increased, road accidents have increased, the average number of miles travelled per vehicle has increased, but the amount of road space available per vehicle has steadily decreased. The roads do not keep pace with the traffic. Until more road space is provided, accidents will increase. Further, that road space must be planned scientifically so that the regulations needed to reduce accidents can be rigidly and yet fairly enforced.

Brains Needed more than Money.

Mr. Rees Jeffreys, chairman of the Roads Improvement Association, claims that the road problem *can* be solved by planning for traffic in advance. The authorities, he argues, tinker with conditions after they have arrived. What is wanted is not so much money, but brains. Brains have never been permitted to handle the traffic problems of the country. Sectional interests and political considerations have swayed the situation. Knowledge and foresight have been divorced from power and money. The Roads Improvement Association has urged the Chancellor of the Exchequer to raise a loan of £50,000,000 on the security of the Road Fund, to be administered by special commissioners, to finance a ten-year programme of new bridges and arterial roads and footpaths. Public support should press forward this plan, for few sounder ways exist of investing some of the money already collected by the nation from road users during the past few years. An improved road system not only benefits health and reduces accidents, but is in itself a development of the country : and even orthodox financiers might agree that while " money " is lying idle this type of capital development is very timely. We should insist that the Road Fund, instead of being depleted, should be increased, for money so devoted will aid national recovery.

London to Cardiff Road.

Typical of some of the new work that might wisely be tackled by a Government seriously concerned for the nation's interests is the proposed London to Cardiff road, a new development road to bridge the Severn near Chepstow and reduce the road mileage between London and Cardiff by 30 miles and between Cardiff and Bristol by 60 miles. Such a road would facilitate the removal of industry from crowded, unhealthy centres and accommodate its southward movement by putting an arterial road through almost undeveloped country between the Gloucester and the Bath roads. It would provide an avenue along which the large surplus mining population in South Wales could travel eastward to find new homes and occupations, supporting themselves in industrial and agricultural settlements in a rich farming country capable of maintaining a much larger population. It would connect the capital cities of England and Wales by a direct road and place Cardiff " on the map ". It would relieve existing country roads and the towns and villages upon them of much heavy and fast traffic, and so avoid the drastic widening of those roads at great expense and perhaps to the detriment of their amenities. It would permit the application, in an area largely free from previous urban development, of modern practice in town planning and road lay-out so as to preserve amenities, prevent ribbon development, and provide for garden cities and satellite towns.

The sponsors of this scheme estimate the cost at £5,000,000, including the Severn Bridge. Spreading the work over five years, about 90 per cent of the annual £1,000,000 would be paid direct for the regular labour of 5,000 men who then would not need to draw about £250,000 in unemployment relief. Additional people would soon be at work building premises and other development work.

This, and other carefully selected arterial roads, such as that planned between London and Southampton— two cities that surely ought to be properly connected —are certainly wise instalments toward a national planned economy. Typical, too, of many useful links that need building or improving all over the country may be mentioned five specific river-crossing projects for London strongly urged by experts. They are : (1) a new bridge at Wandsworth near Hurlingham ; (2) a bridge between Putney and Hammersmith ; (3) a new tunnel at Blackwall ; (4) a new ferry to relieve the Greenwich Ferry (or, perhaps even better, the high level bridge already alluded to, coupled, the writer urges, with completion of the Thames Barrage scheme with which Mr. J. H. O. Bunge has so actively identified himself) ; and (5) the Dartford–Purfleet Tunnel, with suitable approach roads.

Useful Improvement Schemes Everywhere.

Other important schemes such as the Forth and Tay road bridges, in the neighbourhood of Edinburgh and Dundee respectively, and the Dee Embankment Scheme to give greatly improved access between North Wales and Liverpool, can be found by those willing to plan for greater comfort, in all parts of the country. So long as a scheme is sane and practical, money should be no object. By means such as those used by the island of Guernsey a century ago to build its Market Hall, all these and many more physical improvements could soon be achieved. But even within the cramped confines of the present system whereby the private banker is allowed to levy a perpetual toll on industry for his services in providing it with *nothing but a number of neat book entries*, there is no need to despair. American Road Bond methods might have been applied here at any time during the past forty years : they might still be tried. Between 1894 and 1908 the states of New York, Massachusetts, Maryland, and Delaware issued Road Bonds and so expedited the development of their territory. Other States, such as Illinois, North Carolina, and Missouri, realized in and after 1915 that by capitalizing the current motor vehicle taxes they could give their main highway construction programmes a much needed " boost ".

According to Thomas H. Donald, chief of the United States Bureau of Public Roads, 31 out of the 48 States at some time or another after 1894 authorized and issued State Highway and Bridge Bonds. The total reached about $1,000,000,000, but with the motor vehicle expansion consequent upon the improved and great road mileage, the process of retiring the Bonds is now well under way. Mr. MacDonald takes a typical issue, the Illinois issue of $60,000,000 voted in 1918 to bear 4 per cent interest starting in 1922, with retirement starting in 1926 and planned to finish in 1944, and shows that the money realized from the issue enabled 1,480 miles of good paved roads to be completed in 1925. At an average annual payment of $4,052,200 amortization will extinguish the liabilities in 1944, as stated. If the same average annual sum had been used to pay for these roads directly, they would not have been finished until fifteen years instead of five years. In other words, 1,480 miles of road were available for an average period of five years under the bond plan before they would have been available under a pay-as-you-go plan. Mr. Mac-Donald further estimated conservatively that each mile of these roads was used each day of the five years by 1,000 vehicles, and that the owners saved 1½ cents a mile through ability to operate over good-class roads, and so computed a saving in operating costs over the five years of $40,515,000, which much exceeded the interest cost on the bond issue. Of course, other benefits not easily calculable are the quickening of all economic life within the State and such related general benefits. The Bonds are usually of two types, serial or sinking fund, with increased preference for the serial type.

No country, it would seem, has as yet approached completion of its major highway system to the new standards set up by the motor vehicle, so that the problems of highway finance are universal and urgent. Bond issues are feasible and desirable for reasonable construction programmes, but subsequent maintenance should be provided for out of current revenues. Bonds, backed by the State's full credit, though based primarily on user revenues for interest and redemption, should be serial in form with maturities so arranged that annual requirements of principal and interest will be as nearly uniform as practicable.

Abnormal Loads by Road.

Mr. A. R. Polson, F.I.T.A., M.Inst.T., traffic manager to Messrs. C. A. Parsons & Co. Ltd. of Newcastle-on-Tyne, consistently points out the importance of good and adequate roads to British industry, from a fresh angle. He is especially exercised about the transport from works to destination of heavy and bulky pieces of machinery. Certain heavy, valuable articles needed in modern industrial plants and power stations should be of such dimensions that they cannot possibly be conveyed over the British railways : other pieces can only move by rail under special and very costly working conditions. The American, the Canadian and nearly every Continental railway gauge is wider and bigger than ours. The sad result (sad under our present mad system) is that when big overseas contracts for electrical and similar machinery are open for tender, Britain is automatically shut out from competing, or forced to compete under a heavy initial handicap of cost, unless some means of transport other than rail can be used between factory and port. To deal with such " abnormal indivisible loads " one or two haulage contractors and vehicle manufacturers have done remarkable pioneering work, for by designing a multi-wheeled type of vehicle they have succeeded in moving a girder weighing 98 tons on a vehicle and gear of 56 tons weight, on 30 rubber-tyred wheels, the weight in contact with the road being so distributed as not to exceed 3·85 cwt. per square inch. The only known road vehicle previously capable of carrying 98 tons, built some years ago, exerted a pressure on the road of at least 96·63 cwt. per square inch.

Remove Road Level Crossings.

Main arterial roads should take on, in a measure, the character of the railway, by going over or under important cross routes and by-passing intermediate towns and villages so that these connect by means of short branches. Their abuse by the speculative is tardily to be stopped, though much mischief has already been done in ribbon development.

When an arterial road is brought into a regional system it will often be worth while to provide width enough for a central and two side roads (for local traffic), thus reducing the number of crossings on the central road and allowing those crossings to be placed out of line with the openings into side roads. Ample space and clear vision are essential, and are now being provided, at road junctions.

Super-elevation and absence of new blind corners are also now being watched closely in all new road plans ; dual carriageways and roundabout crossings are familiar ; fly-over functions will increasingly have to be adopted ; while successive types of automatic traffic signal lights incorporate fresh improvements. An aim must be to remove all surface tracks for trams or so to widen the roads that the trams can be accommodated on a special track as in the Bristol Road, Birmingham, and some of the outskirts of Liverpool. The Ministry now seems seized with the need to remove the toll system from our roads and bridges, and further progress here should be pressed for. Parkways are overdue in this country, and one hopes that highway authorities will soon be empowered to purchase compulsorily land up to 220 yards from the middle of the road for the purpose of preserving a view or creating a parkway.

How far does Road really compete with Rail?

Ministry of Transport returns for 1933 show that, as against a revenue from passengers of £51,000,000 earned by the railways, and of £6,700,000 by the London Underground, the entire revenue of the road passenger transport industry was £81,100,000. Of this, £22,900,000 was taken by trams and trolley vehicles, and £51,300,000 by omnibuses. £2,600,000 only was taken by motor coaches on regular services, £1,800,000 by coaches on excursions and tours, and £2,500,000 by motor coaches on contract. Now, obviously only some part of the £2,600,000 earned by coach services will really have been competitive with rail. Yet heavy taxation has been imposed, at railway instigation, on the local bus operators who only in a very minor way, if at all, compete with the railway, and so a vital need of the people (the average fare of the 5,335,200,000 passengers carried was 2¼*d.*) is heavy taxed, for no logical reason but to appease railway spite and provide easy revenue to be " lifted " into the general exchequer if desired.

CHAPTER EIGHT

LOOKING AHEAD IN AIR TRANSPORT

" COMMUTERS " now regularly travel by seaplane from Oyster Bay, Long Island, some 25 miles to Wall Street, New York City, their 'planes using a floating pontoon in an East River dock within easy walking distance of Wall Street. The pontoon is anchored at one end to the masonry of the dock wall and may be sunk to water level by compressed air, which medium is also used to operate the forty-five foot turntable when swinging round the 'plane.

Whether such a service could be organized yet for London business men is open to doubt, for the Thames is too narrow a waterway for safe landings and take-offs during all conditions of tide and wind. An amphibian Autogyro, or one designed as a float sea-plane, has not yet been tried ; there would be some risk that rough weather would damage the blades of the rotor. Perhaps Londoners must continue to reach their offices by surface or subterranean means until roof-top aerodromes become the order of the day.

Things are happening in this transport medium, however, far faster than the average man realizes. It was on August 25th 1919 that a small converted war-type British aeroplane instituted the world's first daily air service, for passengers and urgent parcels, between London and Paris. The fare was twenty guineas, and such a flight was an adventure, for no wireless telephony was available and only a quite elementary meteorological service. Soon after, a route to Brussels was opened with big twin-engined bombing machines converted to civil use. In November 1919 the Post Office contracted to employ the London–Paris service for mails surcharged at a fee of 2s. 6d. per letter ! A service to Amsterdam was started in 1920, but in 1921 foreign services with appreciable State subsidies made such a drastic reduction in air fares that the British concerns could not continue without an official subsidy. Such a scheme was instituted, amended and amplified and eventually, on the advice of a Committee appointed by the Secretary of State for Air, Imperial Airways Ltd. was created as a new national company into which existing British lines were merged. It had a capital of £1,000,000 and was guaranteed a subsidy of a like amount, graduated and spread over ten years, but was left entirely free by the Government for ordinary commercial development. March 1924 was the date of this merger, and Sir Eric Geddes directed the new State-aided enterprise. The route mileage at that time was 1,760 and by the end of 1934 it had grown to 20,500. In the first year the company carried 391,000 ton-miles of traffic and in the tenth year 2,733,000 ; an improvement effected by following what Sir Eric Geddes has termed " a steady, plodding, non-spectacular course ".

Penalties of Pioneering.

Some may think this latter remark contains a suspicion of an apology, for of late the air combine's policy has come in for some press criticism. In so

141

far as the criticism is justified, the fact is that we now suffer a bit in the air from having been pioneers. The subsidy paid to Imperial Airways was based on an obsolescence period for aircraft of five years, and except by abnormal expansion of traffic or extension of routes it is not deemed economic to scrap aircraft that are still doing excellent work. Yet such tremendous progress is being made in detailed design that rival companies operating smaller aircraft can take advantage of improvements more easily.

Imperial Airways' history has been one of compromises based on the effort to reach a self-supporting basis through making the public air-minded by judicious publicity and, more important, by providing for comfort and safety from the start. Compromises have had to be reached between range of flight and paying load ; speed and the cost of speed ; reserve of engine power and paying load ; a fleet of larger and more economic aircraft and a numerically larger fleet of smaller air units giving greater elasticity of operation ; and stalling or landing speed, and cruising speed. Illustrating the importance of well-judged compromise, Mr. G. E. Woods Humphery assumes a 'plane requiring 3,000 horse power to propel it at 120 m.p.h. By reducing the speed by 10 m.p.h.— or about 39 minutes longer for a 700 mile run—the pay load can be increased by 750 lb. by a saving in fuel consumed. The value of every pound of pay load has been assessed : it varies with the route, the class of service and conditions of operation, but it may be anything between £5 and £10 per annum, so that the 750 lb. gain in pay load means an increased potential capacity of between £3,750 and £7,500 per annum per 'plane, plus the saving in fuel cost. It will increase, naturally, with the progressively improving mileage output per unit of fleet.

Critics of Imperial Airways stress that though the company does study its passengers, the same cannot be said for the mails, which travel in the same 'planes and are delayed every night while the passengers sleep at an hotel. The business community has long demanded a faster Empire mail service but has not appreciated the difficulty in operating it without adequate ground organization and no increase in direct subsidy. The regular mail to Australia could not be flown night and day with the present ground equipment. The lesson of the Trophy Race, therefore, is surely that the route must be prepared so that the feats of Scott and Black, Parmentier and Moll and others flying all out under dangerous racing conditions, should become possible commercially within a reasonable period of time.

Special Requirements of Commercial Aviation.

This matter of route equipment is one of several which point clearly to the need to separate commercial aviation more adequately from the Air Service. The Air Ministry is more concerned with defence than with commerce, and there is a risk that the Service view on wireless, meteorology, lighting, etc., will carry undue weight and deprive the business side of air transport of the benefit of the latest commercial developments. To take an example from the equipment side. No Service 'planes are equipped with variable-pitch propellers, retractable under-carriages or the most modern devices to secure slow landing speed. Again, the Royal Air Force does not believe in lighting the aerodrome, in a commercial sense:

a Service pilot must learn to land almost in the dark, for he might not be able to be supplied with a flood-light in war time. So the Air Ministry is inclined to whittle down the proper requirements of what is purely a business concern. For night flying to be practised everywhere on our Imperial routes, complete ground aids must be provided. Anything less than the best obtainable, satisfactory to operators and pilots alike, will lead to disaster. The cost of providing these air routes—the permanent way of the air—with signal boxes and staff to man them, must apparently fall on the Government, and so long as that cost is met by reformed financial means, that is as it should be. In any case there is a strong argument for the Government assuming responsibility at this early stage for the complete ground organization of the main air routes.

Aircraft Superior to Weather.

The air cannot be fully conquered while flying is immobilized by bad weather. Side by side with the development of more selective wireless equipment, progress is being made in the production of a completely automatic pilot. One such instrument has taken a 'plane off the ground, flown it to a pre-determined height, depressed the elevators for level flight, at the same time maintaining the course which was set in advance. If height is lost in bumpy weather, the Robot brings the 'plane back to its proper level. The most notable flight yet made with the aid of a Robot pilot was that of Wiley Post round the top of the world. He could sleep whilst the automat kept the 'plane on its course. Corrections for drift must still be made manually, but if the pilot himself can

144

leave the controls, he is free to take bearings or concentrate on instructions received from the ground by wireless. Since night flying in bad weather will become much less difficult when the automatic pilot is a standard fitting, the experimental work should be pushed on.

Even more important is the problem of landing in fog. Though wireless control can now help a pilot to land safely on an aerodrome quite hidden by fog, the needed equipment is still elaborate, personnel need special training to take charge of such an experiment, and the landing area itself must be very large. America, Germany and Holland are tackling this fog-landing question with vigour ; we should be doing so, too, for reliability of operating is vital if the business world generally is to turn to the air as naturally as it to-day turns to road or rail for the reasonably prompt meeting of its transport needs. As Major R. H. Thornton, M.C., M.A., points out, it is useless to the business man to convey him to his destination at 200 m.p.h. on nine days out of ten. Until he is offered reliable transport on ten days out of ten he will not be seriously interested. It is just the ability by the aid of modern instruments to fly without reference to a visible horizon that makes that tenth day a flying day instead of one on which all services are reluctantly cancelled.

Air Traffic Control becoming Urgent.

Experts consider, too, that radical changes in the system of air traffic control are now needed. If aircraft are free to fly under conditions of nil visibility, then aerial collisions will occur with calculable frequency unless the movements of all aircraft are controlled. Any system of air traffic control will demand from aircraft crews a high standard of skill in both radio technique and sound airmanship. With few exceptions, there may then be no place for the amateur pilot in any controlled traffic system and, unless he be extinguished altogether, he must be found a plane of operations where he can do no harm to any but his fellow amateurs. Even with the small traffic and slow speed to-day, on a day of ordinary English drizzle the uncontrolled movement of aircraft over half the county of Kent is regarded as hazardous : while aircraft allowed to ascend or descend through cloud at will are indulging in a gamble. But one cannot visualize the future of air traffic in a thickly populated country such as ours without foreseeing some system of rigorous and unified control.

The present methods of dealing with this problem between London and the Channel ports prohibit air-craft without radio in an area of 1,800 square miles over Kent and part of Sussex from flying in cloud at all, and no aircraft with radio may enter cloud without reporting its position and intended movements and discussing the latter by radio with the Control Officer at Croydon Airport. Within a smaller zone round Croydon itself special rules come into force when visibility is less than 1,000 yards. Into this zone of 600 square miles no aircraft without radio may penetrate without first landing outside and telephoning for instructions ! Aircraft with radio must report their position and proposed movements before entering the zone, and will then be advised whether to proceed or vary their proposed movements, or even loiter out-side the zone till the airport is prepared to receive them.

146

Present Methods Inadequate.

Do these regulations constitute a sound basis on which to develop a system designed to deal with traffic of much greater density? Major Thornton is among those who consider them merely a makeshift, and hold that the principle adopted so far is a fallacy in itself and incapable, anyhow, of large-scale development. The principle pictures the air as comparable with the open sea, an element in which aircraft are navigated, when in or above cloud, by means of radio cross-bearings. Some are converging on a popular airport, others radiating from it, and others just passing in the vicinity. All report their position, height, track and speed to a central control. The Control Officer has to visualize rapidly a constantly shuffling three-dimensional picture of about 1,200 cubic miles of air dotted with aircraft like currants in a cake, moving at varying angles, heights and speeds and all talking at once, while demolishing his picture for him at the rate of three miles a minute.

Even for quite a moderate intensity of traffic, the " open sea " analogy will not work. The total area of the populous part of England is only 40,000 square miles. If the approach to one aerodrome alone is to involve, in poor weather, a continuous committee meeting extending over an area of 1,800 square miles, we need only twenty or so busy airports and Babel is with us once more. Major Thornton argues that we must canalize the whole of the scheduled air line traffic into lanes marked by radio range beacons and divided into " one way " height levels. These levels would be brought into force by an " order of the day " when cloud conditions on any section of the route demanded it. In the vicinity of an airport they would be in constant application. At whatever upper altitudes they might be fixed for the day, they would never be lower than, say, 2,000 feet, except within the approach zone of an airport. In this lower stratum private flying would go on, and in machines that would rightly be regarded as death traps unless capable of maintaining horizontal flight at speeds down to 10 m.p.h.

Greater Volume of Faster Air Traffic Coming.

Nobody knows what the intensity of inland air traffic will soon be, but speeds will certainly greatly increase, and as high-speed transport units of small carrying capacity are developed and service frequency increases, a change amounting to a revolution in our conception of internal traffic becomes possible. One cannot safely assume, then, that flying services between, say, the Midlands and London will never compete with railway expresses.

If this picture is at all correct, some form of organization on a national scale will have to be created with far-reaching powers and duties to deal with the traffic involved. More than the mere provision of aids to navigation and the exercise of a general supervision over traffic in the interests of public safety will be needed. Traffic throughout each national air route will have to be controlled by a homogeneous and unified routine in which all ground personnel have been trained and with which all pilots are familiar. Business opinion does not yet favour placing these duties with a State department, for " its close dependence on Parliament makes it unsuited to assume responsibility for the continuous maintenance of large scale public utilities ". There is, however, increasing support for the timely placing of these big responsibilities with a statutory, self-governing Board such as the B.B.C., L.P.T.B., or C.E.B., and such Board to control air routes was foreshadowed in August 1934 in a memorandum presented to the Government by the London Chamber of Commerce. As then planned, the Board should consist of three or four members, appointed by and responsible to the Secretary of State for Air, with a limited term of office, not less than five years. Within its terms of reference the Board would be a self-governing body corporate, with powers to make by-laws, raise loans and levy dues. It would administer all State-owned civil airports, license all civil aerodromes, equip and administer all ground stations for radio or other communication with civil aircraft; select, equip, and administer airways as required for civil air traffic : license the operating equipment and personnel of aircraft using national airways : collect dues, on a scale approved by the Air Ministry, in respect of aircraft using national airways and radio equipped aircraft : make regulations for controlling traffic on national airways : and make recommendations to the Air Minister for the issue of General Air Navigation Orders.

National Airways Board's Duties.

The Board would be aided by an Advisory Council on which Local Government, Commerce and Industry, Air Line Operators and other interests would be represented. The Board's main function would be the selection, equipment and control of individual specified airways as might appear desirable in its own running survey of the country's needs, and in consultation with operating and consuming interests. These airways would be analogous to the old turnpike, or the modern Italian " autostrada ", a kind of self-governing highway with control points and a code of traffic discipline laid down by by-law. It would be an offence to use the airway other than in an aircraft whose equipment and pilot were licensed by the Board. National Air Dues might be collected by an annual basic tax on radio equipment in aircraft and by dues levied per passenger or unit of freight loaded or discharged at any scheduled airport on a national airway.

Since the Board's operations could not hope to be self-supporting, at least until traffic attained considerable dimensions, it must presumably start on borrowed money (unless, of course, a sensible Government has adopted Social Credit). When mails are carried regularly by air, the Exchequer will probably need to make some direct contribution towards the maintenance of national air highways. Anyhow, the Board might well claim to receive from the Exchequer, as a primary source of revenue, the computed yield of the duty on petrol consumed in aircraft. As applied to air transport this duty is hard to defend, for it becomes a tax, and a very high one, on travel.

It is not too soon to press for this National Airways Board. Civil aviation has already suffered from the vicious circle of " wait and see ". Even to-day some local authorities decline to provide aerodromes until there is an " effective demand " for air transport, yet how can air transport ever become effective without aerodromes ?

One earnestly endorses this plea for a National Airways Board, though not for the same reasons as some who publicly urge it. When private enterprise finds itself in financial difficulties, or gravely puzzled as to its next steps, it hurries at once to the State for help, while continuing to resent any or much interference by the State with the profits when they come. Still, if the demand, however motived and inspired, leads to successful assumption of a more active national interest in aerial transport progress, well and good.

The Aerodromes Advisory Board.

Meantime, a body formed on the initiative of the Royal Institute of British Architects was re-constituted in September 1933 as the Aerodromes Advisory Board, presided over by Capt. the Rt. Hon. F. E. Guest, C.B.E., D.S.O., M.P., and this Board's proposals for a survey and scheme of development for future air routes and aerodromes in Great Britain received the endorsement of the Air Ministry in May 1935. On completion of this survey a full scheme is to be prepared indicating sites for aerodromes to suit the various air lines of the future. The proper control of aerodrome construction and the possible nationalization of the main airports, with powers to deal with the many legal and other such matters arising, are now urgent if we would avoid repetition of the confusion and waste of money occasioned by the want of any central control with regard to railway construction.

It is loosely and broadly urged that every town of 100,000 folk or more should have its aerodrome : and if one sets as an early ideal one aerodrome per 100,000 people one comes into a field of interesting speculation for Greater London alone. A 12-mile radiused circle based on Charing Cross embraces an area of 452 square miles in which lives a population of about eight millions. This area is 50 square miles in excess of the total areas of ten of the largest provincial towns —Birmingham, Liverpool, Manchester, Newcastle-on-Tyne, Sheffield, Leeds, Bristol, Hull, Bradford and Nottingham could all be placed within this area, and their total population is 2,500,000 less than the 8,000,000 resident in this circle. Many of these towns already have established aerodromes. On the ideal of one aerodrome per 100,000 people, London should have no less than 80. One cannot estimate the number of aircraft—especially in private ownership—likely to be in use in the future : but at first the average man will hardly be able to house an aeroplane at his home. Not every car owner can garage his car at home, even. If only one person in a hundred owned an aeroplane there would be 1,000 aeroplanes to be housed at each aerodrome. Hangar space would vary with the type of aircraft, but ample new aerodromes will evidently be needed, and plenty of space round them.

Another aspect of this problem will arise if the autogyro taxi service becomes a commonplace link between the domicile and the main aerodrome, for space for these small craft would have to be pretty considerable even if they were allocated only to one corner of an aerodrome.

Will Airships " Come Back " ?

Though little is now heard in Britain about airships, the Germans are operating with some success over a very long distance, and the group in this country of staunch enthusiasts for the lighter-than-air craft may some day stage a big " come back ". This will further complicate life, for airship sheds involve very difficult and complicated engineering problems, such as those in the very fine adjustment needed when the nose of the airship is being engaged with the top of the mooring mast. Lighting arrangements must be gastight to prevent any chance of an explosion and, to guard against fire when two sheds were close together, it has been necessary to provide a water screen between the sheds, which meant pumping large quantities of water at high pressures.

The author hopes to see, in the future, a body of suitable public utility type which will be the effective owner, on behalf of the whole nation, of all internal airways, and which would co-operate, on some flexible basis, with municipal or other local owners of individual airports. While the older generation doubt the practicability of landing on roofs and flying into the middle of towns, there are enough younger experts, untrammelled by too many fears, who do visualize a time not too far distant when short-distance as well as long-distance flying will be so comparatively easy that the volume of flying will be beyond all comparison with to-day's volume. Short-distance flying may, indeed, help to solve some of the problems of other forms of transport.

CHAPTER NINE

HOW RAILWAYS MAY PULL THEIR WEIGHT

IN a very few years' time British railways will be either unified or nationalized. From some aspects it will not be very material which happens ; but from others vital and distinctive considerations arise. That the present situation is satisfactory to few must surely have been demonstrated earlier, so heed must be taken of the powerful influences making for further changes.

To consider, first, the simpler jump. All the arguments that were advanced to justify the grouping of the railways under the 1921 Act may be adduced with equal force to press the four groups into one. Gradually Parliament has abandoned its early attitude of seeking to retain, in the alleged interests of the public, the competition arising from alternative services. Much of it vanished when railways were grouped, and most of what little genuine competition remained is being discontinued under the pooling agreements. Aided, as necessary, by the Clearing House machinery, the railways more and more tend to speak with one voice and act in harmony on all important matters. Some of the economies visualized under the 1921 Act are now being secured, but the continued existence of the four groups and the small joint and other lines, prevents the full adoption of new economy measures such as would be possible to a unified system. Few will to-day urge that the further drastic reduction of directorates in favour of a few really business-minded full-time experts would be anything but an advantage.

Unification of British Railways.

Pressure for unification may be expected from some elements in the railway world itself: most shareholders will now probably welcome it and the administrative and operating sides will have no quarrel so long as persons rendered redundant are fairly compensated. The trading public's opposition is not now likely to be serious, for they appreciate the position and see that their true safeguard against monopolistic acts lies now with road transport.

But quite likely the railway question will again become an important political issue and socialization on the modern model will be seriously proposed, and, if pressed with any big weight of votes, may be accepted by the railways' financial advisers with reasonable grace as carrying advantages in security of revenue not otherwise to be gained. This idea has lost most of its terror for the public, since experience has shown how the exchange of stock can be achieved with comparative ease and painlessness. And the more the railway propagandists point out how indispensable are railways to the nation, the more do they prove the case for nationalization, or something near it. Should the present political opposition gain power they frankly propose to nationalize railways, and quite possibly the same skilled mind that conducted most of the negotiations with the London passenger transport undertakings will be able fairly quickly to show the main line railway experts the wisdom of accepting an equitable settlement, less worse befall.

True Position of Railway Shareholders.

Responsible people do not now raise the old howl of " confiscation ", either to demand it or to complain of it ; yet should it become a " cry " used by any group of opponents to socialization, it should suffice to remind them of their history as part proprietors of our national system of railways. Railway shareholders were originally granted limited monopolies *on certain conditions* : one laid down very clearly in the Cheap Trains Act of 1844 (7 & 8 Vic., c. 85) permitted the State compulsorily to acquire all railways built after its passage at any time after 1865 on a quite specifically stated basis of purchase. So, every purchaser of railway stock since that date has voluntarily decided so to invest his money at his own risk so far as concerns the permanent continuance of his investment under his own control. (Incidentally, how far does the individual holder of railway stock actually control railway policy to-day? About two-thirds of the 780,000 stockholders own £500 or less stock each. Instead of some 200 railway shareholders' meetings each year there are now but four, and at these, when formal business and votes of thanks are allowed for, there remains about 30 minutes in which the boldest of shareholders may raise his voice for a short space—unless snubbed or ignored by the chair.)

There can be no real grievance, then, so long as reasonable terms are fixed for the transfer of the railways to the nation—and, bear in mind, " reasonable " must mean reasonable to the nation as well as to the railway interests. It is convenient sometimes to forget the pre-War history of our railways and to try to ascribe all their troubles to post-War road competition. Study of pre-War dividends shows that even when railways were operating with the greatest freedom the public would permit—were, in their way, dictators—and were paying very low wages and salaries, complaints by traders and shareholders were almost uninterrupted, dividends were seemingly not good (though " watered " capital partly explains this) and the systems naturally seethed with unrest.

Only after determined agitation by fiercely dissatisfied shareholders, stubbornly resisted as long as possible by its Board, did the S.E. & C.R. permit the addition to their directorate of one individual (carefully chosen by themselves) with expert experience on another railway. Between 1902 and 1910 the South-Eastern Railway Ordinary stocks paid from 2 per cent to 3½ per cent—mostly 2¼ per cent—and London & South-Western Deferred Ordinary paid from 1½ per cent to 2⅛, mostly 2 per cent or under. The Great Eastern Railway averaged 3¼ per cent and the Midland Railway 2¾ per cent on ordinary stock, though the London & North-Western, London, Brighton & South Coast and North-Eastern paid somewhat better dividends upon stocks the market price of which, however, was mostly well above par, so that the yield to the current purchaser was seldom more than about 4 per cent.

Give them a Fair Deal—but No More!

Railway shareholders would be fairly dealt with if they got, in exchange for their share certificates, new stock based on the average market prices of their holdings over several recent years. The " widows and orphans " cry has been raised in railway stock matters so often that it defeats its own purpose. Railway shareholders are perhaps unfortunate in not having found means of impressing their wishes more effectively upon their servants the directors, but they are not alone in that unhappy position, and there are many other unfortunate features in the social set-up. On the other hand, they have had excellent political representation. Railway shareholders deserve justice, but not preferential treatment over all other industrial shareholders and over the canal and coaching interests whose ventures were put out of business by the railways. By doing them justice, and no more, in the transfer of their property to national ownership with, probably, commercially minded direction, the first big burden on the properties might be eased.

Parenthetically, one or two reminders may be given before the argument proceeds. Of the £1,300,000,000 of railway capital on the books at the end of 1913, £198,000,000 or 15 per cent was shown by the Board of Trade as being " water ". In 1844 Parliament enacted that there could be a revision of railway charges if the dividend of any railway was more than 10 per cent. This may possibly explain " watering ". For instance, the Taff Vale Railway Co. gave every holder of £100 preference stock £125 of new 4 per cent preference stock and £150 of ordinary stock ; and every holder of £100 ordinary stock received £250 of new ordinary stock. Subsequent dividends on the latter at 4 per cent sounded reasonable, especially when addressed to the workers in justification of lower wages. Remembering such facts, and relating recent dividend rates to recent market prices of railway shares, there does not seem to be so much in shareholders' complaints as the public has been led to believe.

Sir Eric Geddes' Reorganization Plans.

After the War, Mr. Churchill announced that railways would be nationalized, but the Coalition Government returned them to company management, with a gift of £60,000,000 of public money, and gave Sir Eric Geddes, first Minister of Transport, the task of reoganizing the companies and eliminating the worst features of their wasteful and overlapping methods. As he said in Parliament, " You must make one block of capital do the work now, not two." An important item in his programme was abolition of the 650,000 privately-owned wagons, which he said would yield a saving of over 20 per cent.

As usual, private interests prevailed, and the Coalition Government perpetuated most of the old evils. Some are preposterous. Several years ago the South-Eastern Railway obtained running powers into Eastbourne, after spending a large sum on legal and other fees. When it obtained these powers, the Brighton Company paid the South-Eastern £38,017 every year *not* to run into Eastbourne. The public now pays this sum over, apparently, for it was included in the 1913 profits of the South-Eastern, and was counted as a working expense before the Brighton Company arrived at its profit. These two companies are now merged, and as their fares and rates are based on the 1913 profits of each, every passenger or consignment of goods carried on that system pays some share of this perpetual tribute ; and the 1921 Railways Act entitles any group to apply for power to raise its charges should it fail to earn its standard revenue, i.e. its 1913

profits plus allowances for capital since supposed to
have become interest-earning.

Alternative Basis for Railway Socialization.

An objection to transferring the railways to the
nation at current or average current market value
might be the risk of manipulation of the market against
the public interest, directly definite nationalization
plans were made known. If it were not found feasible
to guard against this risk, a plan of almost equal merit
would be to adopt the fair basis of reasonable net
maintainable revenue. This formula is the one which
Mr. Herbert Morrison sought, without success, to get
through the Joint Select Committee of Lords and
Commons on the London Passenger Transport Bill.
Instead of the actual sums payable being settled by a
small body of arbitrators whose personal opinions
and susceptibility to the skilled advocacy of the parties
might have some influence on the result, the sums should
be specifically set out in the statute or other instrument
which effected the socialization. Thus, while Ministers
could negotiate and argue with the railway interests
as to the basis and its detailed application, they and
Parliament would have the last word ; a useful factor
conducing to voluntary agreements.

Critics of this idea should remember, as Mr. Herbert
Morrison has shrewdly pointed out, that Parliament
determines the amount of compensation to be paid
to wage earners for loss of wages during periods of
unemployment ; they also should recall that the
railway companies strongly objected to the principle
of compensation at all in the case of workers displaced
under the pooling schemes mutually arranged between
the railways ! Evidently the financier's mind has

strict standards in the matter of compensation for the working classes, even respecting compensation for industrial accidents under the Workmen's Compensation Acts, long resisted by employers ! So conservative-minded capitalists cannot object if reasonably strict safeguards for the public are imposed when private undertakings are to be compensated on being socialized.

Skilled Commercial Management Continued.

There should be no State guarantee of interest, but under commercial management by a few skilled railway commissioners in frequent consultation with the National Transport Board and working, virtually, as a sub-committee of that Board one need not imagine that the results to the present proprietors of the railways would not, at the very worst, be markedly better than anything they have known of recent years at any rate. For the way would soon be clear for material fresh economies in managing and operating costs concurrent with the continued steady campaign for greater real efficiency and better service which already is under way.

Among the useless lumber to be thrown aside in the measure implementing this great change would be the whole conception of the Standard Revenue, which has handicapped the thought and action both of the railways and of the Railway Rates Tribunal since 1928, while serving no compensatory purpose but to increase the fees of a few highly-skilled legal advocates when the Tribunal makes its Annual Review of Railway Charges. Part of the theory that led to the plan —that of the division among the public of railway profits above a stated percentage—was sound enough when conceived ; but traffic changes have never yet permitted it to take effect, and its purpose will much better be met under a new set of financial provisions which must be worked out in the railway socialization Bill.

For the railways have always been gravely burdened by their load of uneconomic capital which, once raised, has never been repaid even when some companies were distributing quite good dividends and creating bonus shares. Why railway capital should differ from other capital in this regard has never properly been explained. Probably it cannot be explained ; but whatever the conditions of its original issue, Parliament can alter them if it wills, and, remembering again that every stockholder in railways accepted, when purchasing, the risk of the impermanence of his holding, the State must insist on the setting up by the new undertaking of a sinking fund so that capital shall be gradually repaid, as is being done by the Port of London Authority.

Even those friendly critics of railway economics, Messrs. Kirkaldy and Evans,[1] find that our railways

> have been too prone to borrow fresh capital for extensions, while the ordinary limited liability companies contrive to provide for much more of their growth out of reserve funds. In recent years the companies have exercised more restraint under the compulsion of the expressed opinions of their critics, many of whom have been shareholders, but previously their attitude in this matter was well summed up by Sir Richard Moon, a former chairman of the old London & North-Western Railway Company, who once said, " If the capital account were closed the company would never pay another dividend."

[1] *History and Economics of Transport*, Pitman, 5th edition, 1931.

Better Service from Unified Railways.

But would either unification or socialization of the railways really improve their earning capacity and their capacity to render the community a better and more suitable service and in time, maybe, a more adequate service? This, of course, is the crux of the argument, and, unless the response were a definite " Yes ", there would be no point in taking the discussion thus far. Never forget the irony of the fact that, our engineers tell us, carriage by means of vehicles on metal wheels hauled by steam power over a steel rail is an inherently cheap means of transport, being about ten times cheaper per ton-mile than road haulage even in modern multi-wheeled units. Why is the railways' enormous mechanical advantage lost in current practice in Britain? The top-heavy financial structure, a legacy of past robbery condoned and perpetuated, has been dealt with already. What of the working methods and the mentality? Both are changing for the better, but progress must be hastened. A much simpler basis of charging must be adopted both for goods and passenger traffic. It is already coming into existence for goods traffic, but in the hurry to give desired concessions to the big man—who can now ignore classification and normal accounting routine entirely, if he wishes—the small man who does not send much traffic, but of whom there are very many, must not be forgotten. Either drastic simplification of classification must come quickly, or the railways must actively encourage the setting up of a " spediteur " system like that known on the Continent, to group the " smalls " and offer the railways full truck loads, while sharing the resultant economies with the traders. But why need the railways leave that task to some other interests? Allied with them are now some of the leading parcel-carrying companies in the country; for them to improve, cheapen, and popularize their " smalls " forwarding arrangements would be an easy matter, given willingness.

No railway officer will seriously contend that the freight services are yet worked at anything near maximum efficiency. The number of goods and mineral wagons owned by our railways decreased from 724,380 in 1920 to 682,759 in 1931, while the total tonnage capacity of the wagons rose during the same period from 7,435,460 tons to 7,594,513 tons. This means that the average capacity per wagon had risen in ten years from 10·236 tons to 11·351 tons. Good, though not good enough. But the average haul of goods in classes 7 upwards was 102 miles in 1933, and the average wagon load was 2·83 tons; while for minerals and goods in classes 1 to 6 the figures were 64 miles and 9·15 tons; and for coal, coke and patent fuel 42 miles and 9·40 tons. The average freight train load was 122 tons and the number of wagons per train was 34·12, of which 11·15 were empties.

More Effective Use of Rolling Stock.

This is partly explained by unbalanced freight movements and the need for special types of vehicles for different purposes, but there is room here for more skilful use of stock and a lessening of the heavy proportion of empty running. The exceptional rates system is conceived to encourage good wagon loading, but a system of general percentage reductions for ten-ton lots and five-ton lots, if properly advertised, would give traders further impetus to co-operate for improved wagon loadings and probably press backward traders to change their marketing policy. While it is difficult to quantify the possible savings under this head, or say how far larger goods wagons and their more skilful use will come, yet concurrent progress will doubtless be made in rationalizing industry and agriculture, and more skilfully applying industrial transport management principles to goods movement, so that the railway administration will meet with increasingly active co-operation in working its goods services more efficiently.

Railway reformers periodically have called for larger wagons and the abolition of privately owned wagons. These private wagons convey about two thirds of the coal traffic ; their capacity is mostly between 10 and 12 tons, which is less than half that of Continental coal wagons. Sir Ralph Wedgwood, Chief General Manager of the London & North-Eastern Railway Co., calculated that the elimination of the private wagons would mean a saving to the railways of nearly £1,000,000, but " the great bulk of operating opinion is inclined to the view that the savings effected would be vastly greater " (*The Times*, May 16th, 1932).

Every railwayman would like to use the largest wagons he could, for one twenty-ton vehicle costs less to build and maintain than two ten-tonners, occupies a smaller space on the running lines and in sidings, and requires less locomotive power for its haulage. Though the rapid transit of " smalls " is always trotted out as the difficulty here, and nobody disputes its desirability, there remains the bulk of the tonnage of heavy and low-class material which could move in larger wagons if the whole problem of its movement were tackled in a commercial spirit by traders and railway officers. One recalls the " alibi " that larger wagons would be useless to many collieries without costly alterations at the pithead, and the very partial remedy of this difficulty when finance was forthcoming for the purpose. Yet colliery amalgamations leading to the pooling of their private wagons and the availability of more finance for useful employment, plus a more generous sharing by the railway administration of the benefits secured from using more high capacity wagons, ought to effect big improvements here. If they cannot, then the drastic re-siting of industry and the increased transfer of such coal as is then carried to other means of transport are likely ; but one can surely look to a national administration to sort out these complex factors into their proper relationship and solve the problem which space prevents being further treated here.

Many Openings for Further Economies.

By way of suggesting other economies, note that since the amalgamations under the 1921 Act the number of manufacturing and repair plants owned by the railways have been reduced to ten, and their productive capacity is still greater than demand warrants. Despite constant pressure from manufacturing interests, railways continue to build their own rolling stock and equipment. Further savings could certainly be made by unifying this work.

Another simple direction in which a unified administration would save money lies in the better organization of local cartage services. Already this is proceeding as part of the benefits of pooling, but it can only be perfected under unification. Need four highly-organized collection and delivery services cover the London streets, so that lorries for different railways follow each other into traders' despatch decks? Admittedly the fullest benefits of a unified collection and delivery service can only be gained when a centralized clearing house, an adaptation of the scheme so earnestly advocated by the late Mr. A. W. Gattie (a pioneer who had a very poor deal at the hands of the contemporary railway mandarins) is actually in being. Some American cities have now shown us how to run a " Union Freight Terminal ", and there is no reason why London and other cities should not adopt the practice. Railway containers, introduced at traders' pressure and now sworn by in the railway service, are tending to make poor Gattie's ideas more acceptable to railway officers, and an early duty of a unified administration would be to re-examine the scheme and test it in a small way in some city of moderate size.

The Very Controversial " Green Arrow " Service.

One wants in this chapter to keep to fundamental issues, or much could be said on many minor grievances of goods traffic interests such as, for example, the collection of many thousands of half-crowns for " Green Arrow " specially expedited service when, in fact, nothing is done in most cases that is not held out as being the regular thing when " next morning delivery " in almost every part of the country is the normal motto and advertised claim.

Abolish First-Class Travel.

But though economy and efficiency in the goods services are vital and will be of value to the general revenue and the goodwill of the new railway administration, it is its achievements in making the passenger welcome and getting him to his destination more quickly and comfortably that will naturally appeal most to the general public. And here my first material suggestion may seem, at first, a step back. The time has come, I submit, to abolish first-class coaches almost entirely. First-class travel has persistently declined since 1913. Ministry of Transport returns show that for 1932, while first-class seats represented 13 per cent of the total seats in passenger trains, the percentage of first-class journeys to total journeys was only 2 per cent—ranging from 2·39 per cent on the Southern Railway to 1·08 on the Great Western. So that a very disproportionate number of seats are being reserved and a heavy waste tonnage is being moved about and adding to train running, lighting and staffing costs, beside original construction and maintenance costs. Worried by this problem, the railways recently made a small cut in first-class fares—which is tackling things the wrong way. Most tube railways operate quite successfully on the " one class only " principle, and that class, though pleasantly seated in comfortable modernistic coaches, does not have lavished on it the inordinate waste of upholstery typical of our railways' first-class coaches. To raise the level of the " third class " rapidly to modern ideals, and add Pullman, Restaurant or other *de luxe* cars only where needed, is the real solution ; it being of course understood that cars wherein refreshments would be served would come into still wider use as part of the normal travel amenity on practically all long-distance trains.

Travellers who used to insist on some distinctive comfort and on separation from the " common herd " now mostly use their own cars, and many will soon use aircraft ; so that on the odd occasions when they deign to use the train they can very well take it as they find it, and they will find it, after a few more years of constructional progress, nothing to be ashamed of.

Simplified Fare Suggestions.

How far simplification of fares and travel conditions will go would form a subject for much interesting conjecture, for one must visualize a great easing of the peak load problem partly by judicious " stepping " of hours of labour by employers, partly by better housing of workers in relation to their tasks, and partly by the greater share of freedom enjoyed by each citizen enabling him to spread his travels more evenly over the hours and days. The replacement, even in part, of the present " peaks " and " valleys " by a steady number of good loads would do much to reduce working costs, and a day may come when there will only be three fares charged throughout Great Britain—say sixpence for any journey up to 25 miles, a shilling for a journey between 25 and 100 miles, and half a crown for any lengthier journey, as has been seriously suggested by a prominent engineer. If anybody thinks this preposterous let them remember that a century ago the idea of a letter travelling from London to Aberdeen for the same charge as one sent from the Bank to Ludgate Circus would have seemed equally absurd. Total volume of traffic, simplified accountancy, and relation of terminal to conveyance costs, are the economic factors to be considered in both cases. Meantime, one may note that in 1911 (a convenient pre-War year), leaving season-ticket holders out of account but lumping together first-, second- and third-class passengers, the average fare paid by the 1,326,317,000 passengers carried on the United Kingdom railways worked out at less than 7¼d. What an enormous printing, clerical and inspection saving would accrue if that same aggregate of passenger revenue were earned by the three simple zone charges.

The fear that such system would encourage frivolous travel and so inordinately increase the railway costs hardly deserves serious answer. It is disposed of, firstly, on the analogy of the girl worker in the chocolate factory who, being permitted to eat as many chocolates as she wishes *at work*, quickly loses any desire to be greedy ; and secondly, by asking who is the arbiter on what is " frivolous travel ", and whether the ability of the average Londoner to visit Edinburgh, Aberystwyth, Cornwall, Bournemouth, the Yorkshire Moors and Brighton all in one year, if he wishes, is really a thing to be encouraged or discouraged both from the Londoner's own viewpoint and from that of the inhabitant of the other places named. If greatly increased mobility conduces to better health, wider sympathies, and a truer education, are not the railways which make this possible doing a really nationally useful job ? People appalled at this prospect are suffering from the old slavery complex and must not be heeded.

Rationalized Services.

A national railway administration properly co-ordinated with other transport arms could proceed faster than now with the closing of unneeded branch lines and little-used intermediate stations and the substitution of comfortable coach services making quick and convenient connections at much improved

" transport centres " (not just railway stations) with the fast trains. Awkward cross-country journeys will either be left to the good, reliable coach services which in some cases meet the need, or will be made by modern pneumatic-tyred road-trailers of the " Micheline " type or one of the diesel electric railcars. By using such vehicles wherever loads tend to be light a more frequent service than of yore can be given.

A recent writer, in a quite mildly worded but telling summary of railway failings, said, " Perhaps the most symptomatic feature of all is the average railway station." Why need heaviness, gloom and dirt still exist in any one of them ? Why should waiting-rooms be gloomy cells to which the draughty platform is preferable ? Why should not modern refreshment facilities be counted on with warrant at all leading stations, at the least ? If the railway station were truly the key to the city, what horrible impressions the newly arrived traveller would still have of London and almost every other big city to-day ! Allowing full marks for the work already done in this regard, much more is needed and should be set in hand at once.

CHAPTER TEN

VASTER POTENTIALITIES FOR OUR CANALS

SOME see signs that the younger generation of transport officers are not content to allow our vast system of canals to go to waste. Methods of working are being investigated, new ideas of operation are being tried, and improved appliances and machinery are being installed. Canals, clearly, differ from roads and railways in that the wear and tear of a canal is not commensurate with the traffic passing ; indeed, canals tend to deteriorate faster with the absence of traffic than when in full use. The waterway itself is not subject to wear except for the wash in the banks. The cost of maintenance of a canal is not, therefore, commensurate with the volume of traffic, except for water used in locking up and down, and the wear and tear of lock gates and gear ; other items of maintenance, such as bridges, banks, dredging, fences, general maintenance, rates and taxes, go on independently of the volume of traffic passing. So that the overheads, generally speaking, of the waterway itself are constant. Traffic handling costs vary more directly with the quantities actually passing : the number of boats utilized, the wages paid, the power required, are definitely related to the tonnage carried.

Canal transport is not quasi-monopolistic, as is

railway transport. Since canals can be used by any trader with a canal boat of suitable dimensions and build, any improvement of our waterways must make them better assets to the nation. That is reason enough for giving this subject more helpful attention than it generally receives, and for keeping a much-needed level head when various partisan statements are released. Again, traffic by canal is silent ; something like 750,000 tons pass annually through the heart of London, via the Regent's Canal, without the slightest noise or inconvenience to the public. Could this be done noiselessly by rail or road ? Imagine the increased congestion and danger of London's streets with this extra volume of tonnage thrust upon them. Think, too, of the possibility that great traffics now moving by land might better use the water route.

Traffic Possibilities of our Canals.

According to the Chairman of the Grand Union Canal Co., most British canals could carry ten times the amount of goods now passing over them without any congestion. In typical recent years the canals of Great Britain carried the following tonnages of goods :

1920	.	.	16,305,281	1932	.	.	11,382,922
1925	.	.	15,570,941	1933	.	.	11,434,504
1930	.	.	13,235,784				

The era of canal construction was from 1760 to about 1830, at which latter date our canal system as it exists to-day, except for the Manchester Ship Canal and a few minor branches, had been completed. Most of these canals were at first highly profitable concerns, but as railways developed their flourishing days began to wane. Canals, doubtless, had practically the monopoly of the goods traffic in the areas they served, and although their advent had meant that transport costs generally were reduced, after they had secured the traffic, some concerns tended to abuse their monopoly.

Central Waterway Board Advocated.

During the past thirty years three Commissions and Committees have studied and reported on our canal systems. The Royal Commission on Canals and Waterways, appointed in 1906, issued its final report in December 1909, and advocated the creation by Act of Parliament of a Central Waterway Board, with power to acquire and administer such canals as it deemed expedient. The first waterways for acquirement were suggested to be those comprising the four main routes forming " the Cross " and connecting London, Hull, the Mersey and the Bristol Channel with the Midlands, and a programme of reconstruction was to be prepared and carried out by the Board with a view to reviving their utility and adapting them to modern needs.

Then, after the War, it having become apparent that the difficulties of the canals were even greater than before 1914, a Departmental Committee on Inland Waterways was set up in 1920 by the then Minister of Transport under the Chairmanship of Mr. Neville Chamberlain, M.P. In this Committee's second interim report (June 1921) the conclusions of the Royal Commission were deemed impracticable, chiefly owing to post-War increases in costs, and the following proposals were submitted : (1) the amalgamation of various undertakings into a series of groups ; (2) these groups should be owned and administered by public Trusts, as it seemed impossible to leave them purely to private enterprise ; (3) the transfer, on terms, to the Trusts by the railways of any railway-owned or controlled canal.

Royal Commission's Modified Proposals.

The Report of the Royal Commission on Transport, presided over by Sir Arthur Griffith-Boscawen, was in many ways similar to the two preceding Commissions, but took cognisance of changing events thus :

> Nevertheless, we are of the opinion that certain canals still possess considerable value as a means of transport, and that, properly rationalized and developed, they can be made to render much useful service to the community in the future.
>
> We are satisfied that a process of amalgamation is a necessary preliminary to any development programme. The fact that the voluntary amalgamation which resulted in the formation of the Grand Union Canal Co. has been achieved, and that it promises to be attended with success encourages us to hope that similar voluntary amalgamation schemes will be effected by other canal companies.

The main obstacle to further such voluntary amalgamations, in the Royal Commission's opinion, was the unwillingness of the railway companies owning or controlling certain sections of through canal routes to amalgamate with independent canal companies. It would be no hardship if, where acquisition of a section of railway-owned canal is necessary to the formation of a through route under an amalgamated company, the railway company should be heard only as to price.

Public Trusts to acquire Canals.

So essential were amalgamations deemed to be that the Royal Commission added a further recommendation, often conveniently forgotten by writers on this

175

subject, that the Minister of Transport should take steps to set up public Trusts to acquire such canals as he deemed it in the nation's interests to preserve and improve. The " Cross " routes might be first amalgamated under four Trusts, and ultimately fused into one large group.

Much could be done, the Royal Commission found, to simplify the schedules of charges for carriage of goods on canals.

Unfortunately, the defeatist attitude which seized the canal companies when railways came has very largely persisted to this day. A few concerns are rousing themselves and putting up an excellent fight. Many new factories have been placed on the Old Grand Junction Canal in Middlesex, because traders can get goods as quickly and cheaper from the London docks by water as by road. The railways themselves are developing delivery by canal in the Black Country, and have equipped and extended interchange wharves. A wide range of goods—by no means only low grade heavy traffic—already moves by water, and there are still greater possibilities with better terminal warehouses and facilities, the running of boats by night, and the closer co-ordination of road and water.

Economic Reasons for using Canals.

Turning again to general economic reasons for using canals more. The function of a canal is intermediate between that of a road and a railway. The amount of traffic passing can be much greater per mile than on a railway. Probably the actual cost per ton-mile is lower than by any other method of transport, and nothing can approach the coal trains for cheapness and efficiency. (One refers to the " compartment boat" system developed on the Aire and Calder Navigation by which long trains of specially shaped barges loaded with coal move down to Goole for shipment, and each barge is there lifted very simply and its load tipped direct into the hold of the vessel.)

Some alert industrialists support canals because the existence of a canal between given points has in the past been an influence toward lowering and keeping low the railway rates between those points. Indeed, " water competition " used to be the strong reason for pressing for an exceptional railway rate just as is " road competition " more recently. Railway rates between Birmingham and the Bristol Channel ports still clearly reflect the existence of this water competition, as can be seen if rates for like goods over like distances in other directions from Birmingham are examined by the trained industrial traffic manager. Canal enthusiasts look jealously on the money spent on road construction, arguing that if the Minister of Transport would examine the case for moderate reconstruction of the waterways, and make a percentage grant where a guarantee of interest is not sufficient, valuable national economies would soon be achieved. Curves could be eased and bottle-necks cut out, so that at no great cost many waterways could be made more efficient.

Vigorous Enterprise of Grand Union Canal Co.

What the " Grand Union " have done certainly gives hope to those who maintain that waterways are not yet " played out ", but might be made to serve very useful purposes. The Grand Union Canal Co. was created on 1st January 1929 by the amalgamation of the Grand Junction and Regent's canals and the purchase and absorption of four smaller waterways in the Midlands.

The Canal is entered from the Thames at two points —Limehouse and Brentford—and now operates a through route terminating in the heart of the City of Birmingham; a length of canal which, in fact, forms the complete Route No. 1 of the " Cross " as recommended by the 1906 Royal Commission. Further, at Norton Junction in Northants the canal proceeds in a north-easterly direction to Leicester and the River Trent, and forms part of Route No. 2 of the " Cross ".

Having acquired these undertakings the Grand Union Canal Co. seriously applied itself to the problems of improving the standard of craft and improving the canals.

Between Braunston and Birmingham the company found themselves possessors of a narrow canal with 52 locks for boats of only 7 feet beam, while on the Grand Junction section there were 51 bridges incapable of passing a 14 feet barge, as well as 47 bridge sills and 53 lock sills with less than 5 feet depth of water. The sills did not daunt the company, but 51 new bridges were too much to start on and so the 14 feet beam craft had to be the ultimate aim. The company kept to its intention to provide for a draught of 4 feet 6 inches, but at first this was limited to the through route to pass craft of 12 feet 6 inches beam. Even so, 14 bridges had to be reconstructed.

The cost of reconstructing locks to improved standard, reconstructing public road and accommodation bridges, lowering of bridge sills, culverts, erection of pumping stations, dredging to improved standard, and concrete walling and piling was estimated at £881,000 and the Government made a grant under the Development

(Loans Guarantees and Grants) Act, 1929, of the sum of £500,000 being full interest for ten years and 2½ per cent for a further five years. The reconstruction gave opportunity for the introduction of several new features. Adjustment of the water level in the lock chamber was hastened by the construction of large sluices of a spiral type with bevel gearing specially designed; and the sluice culverts were carried along the side walls and provided with three lateral openings through which the water finds an exit. These large sluiceways ensure the rapid filling or emptying of the lock, and the average time for craft passing through does not exceed four minutes for the complete operation.

Saving Water at the Higher Points.

The saving of water in lockage was an important consideration at Knowle and Bascote, where side ponds were provided for economizing water. Two were provided on different levels at each lock, and about half a lock of water is saved with each operation, i.e. half a lock is drawn from the side ponds, the remaining half being taken from the pond above the lock.

Then again, the extreme variation (from 60 feet to 33 feet) in the top width of the canal made the problem of the most economical relation between the respective cross sections of the waterway and the submerged portion of the barge of great importance.

French canal engineers had studied this matter, but more information was desired, and so experiments were carried out on the Grand Union Canal to ascertain the economic minimum section of canal to be provided to accord with the company's scheme of improvements. Very interesting results were obtained, e.g. for a 14 feet horse-drawn barge, where the ratio of canal area to boat immersed area was increased by passing from an undredged section to a dredged section by 56·6 per cent, the speed was increased by 66·8 per cent ; for a 7 foot motor-driven boat passing similarly into a dredged section where the area ratio was increased by 77·7 per cent the speed was increased by 59 per cent or from 2·73 to 4·62 m.p.h. The result of these tests was that the standard to be aimed at was, for the 14 feet beam craft, a minimum bottom width of canal of 32 feet by 5 feet 6 inches minimum depth, and for the 12 feet 6 inches craft a minimum bottom width of canal of 27 feet.

A new type of grab dredger operated by a 30 h.p. Dorman diesel engine was perfected for work on the Grand Union property, but later a still more efficient steam dredger with the necessary hopper boats and discharging plant were obtained, and from January 1930 to the end of 1934, 831,101 tons of material was removed from the canal.

Without doubt the internal combustion engine has opened up a new era for canal transport, and the horse is fast disappearing. For short hauls the horse still serves a useful purpose, and will probably only give way to some form of electric traction for dense local traffic.

Why not use Canals for Pleasure ?

Sections of canal that do not fit into any improvement scheme, and sections which the changing trends of trade have left comparatively deserted, might still be of national value if considered from the angle of open-air travel ? With judicious invitation and a simple organization to smooth over some of the apparent difficulties of the adventure, many thousands of people would surely respond to the suggestion of an unconventional holiday cruising over the quieter and lesser known sections of our waterways. Small craft can get almost anywhere in England by canal and river, and while the yachting attractions of the Norfolk Broads are well known, some of the beautiful stretches of canal encountered in the West Country and about the Welsh border deserve more attention. Already in the latter case some concerns do quite well by providing barges specially fitted up for tourist sightseers. A small out-board motor launch proceeding only at a leisurely speed and carrying a canvas cover and the usual boat's stores, with a pair of oars, a boat-hook and plenty of rope fender are needed. Speed, of course, will not be sought ; in any case, locks will prevent it. Canal cruising must be a leisurely occupation ; it will not be expensive, although the toll charges must be remembered.

Tunnels will present no difficulty to a self-propelled boat with a good headlight, though they are something of a bugbear to the crews of horse-drawn barges, for there is no towpath through the tunnel. While the horse is taken overland the barge must be propelled through the tunnel by the crew lying on their backs and kicking against the walls—" legging through ".

Is Private Enterprise longer feasible for Canals?

Readers who have heard canals so often spoken of as " back numbers " should remember that this country has never really enjoyed the full economic value of water transport owing to the exceptional circumstances which prevented co-ordination in the interests of through traffic movement. Thus the problem is most emphatically one to be considered as a national question of preserving and using in its due sphere a form of transport that has many natural advantages.

Belgium's co-ordinated transport services include the inland waterways. In this country the striking success of a collectivist undertaking, the Manchester Ship Canal, contrasts conspicuously with the general failure of private enterprise to make the most of Britain's inland waterways. Canal potentialities can only be developed to the full with the nation's aid, and, if that be so, why not under the charge of a national transport authority?

If the validity of the Manchester Ship Canal example be questioned, since for many years the shareholders received no dividend on their investments, there are two crushing answers. First, one offers no defence at all for the working of an insane financial system ; however the bookkeepers juggle with figures the practical value of that project is now abundantly admitted. Secondly, and stepping back for the nonce from that safe position, I submit that many who know that the M.S.C. Co. paid no dividend for many years probably do *not* know that directly the canal was finished the railway rate for cotton from Liverpool to Manchester was reduced from 9s. to 7s. per ton, and on cotton goods from Manchester to Liverpool the rate was reduced from 10s. to 8s. per ton. If the community, as a whole, did not secure this benefit from its enterprise, whose fault was that?

German Waterways used in Nation's Interests.

The extensive system of canals and waterways in Germany, like the docks and harbours, all belong to public authorities, and are not operated for purposes of profit. Towards the end of the nineteenth century and up to 1914 Germany made stupendous efforts to develop her industry and extend her export trade. Her chief weapon in this campaign, without which science and organization would not have succeeded, was the utilization to the full of her magnificent transport system, including inland waterways. Her great toll-free rivers were national highways of commerce, with docks and harbours far inland : at Frankfurt, Strassburg, Mannheim, Dusseldorf and other cities.

The superficial critic is apt to point out that there can be no comparison between the British canals and the State-managed canals and canalized rivers of France, Germany and Belgium, both because all the latter are state supported and because they have enormous natural advantages over our systems in that they have longer level stretches and, in proportion, far less locks than we have in Britain. Precisely ; and if under those advantageous conditions the State saw fit to exploit these natural advantages to the full, how much more does it behove the British Government to tackle still more energetically the problem of further assisting British waterways in the interests not of a select group of shareholders but of the whole nation ? Continental nations would hardly spend so largely upon canals unless they were economically sound. One cannot imagine any Continental country possessing such rivers as the Thames and Severn permitting them to remain in their present unsatisfactory state.

Canal and Railway Directors in Agreement.

It is fairly widely known that in June 1933 the Canal Association (representing the independently owned

183

canals) entered into an agreement with the railway companies for the co-ordination of traffic. The scheme, which arose out of discussions during the Parliamentary proceedings in connection with the Road and Rail Traffic Bill (later the 1933 Act), provided for the setting up of a central conference between the railway goods managers, Canal Association representatives, and representatives of recognized Associations of bye-traders, to promote the greatest possible measure of co-ordination between the two sides, protecting the interests of each other in regard to the quotation of rates for competitive traffics, eliminating undue competition and obviating reduction of rates to an uneconomic level. The canal interests have undertaken that if canal traffic is destined for, or originates at, a point not reached by canal or by services provided by canal undertakers, they will try to influence the traffic so as to get the balance of the task to the railways in preference to road hauliers. The railways have signed a provision in the converse sense.

One more significant pointer to keep in mind.

CHAPTER ELEVEN

USING PORTS IN THE NATION'S INTERESTS

PROBABLY in no single sphere of national life can a more forceful case be made for national planning than in that of port ownership and administration. To-day's port situation reveals a total absence of planning on a national scale and, in consequence, there is going on, week by week, a grave waste of resources which might and should be boldly stopped.

Round our comparatively small coast-line Turnbull lists 542 ports and harbours, and Board of Trade blue books list 171 ports, of which number about 120 have material importance. The Royal Commission on Transport, in their final report, summarized the ownership and management of the docks and harbours of Great Britain thus :

Local commissions or trusts not working for profit, established under statutory authority . .	110
Municipal authorities	70
Railway companies	50
Harbour companies or individuals . . .	100

Like Topsy, ports in the past have "just growed" through the exertion of groups of citizens or of financial interests in a particular district, and the urge of dividend-producing or local patriotism and zealous civic effort have between them accounted for much of the progress on which we can, up to a point, congratulate ourselves. Fair enough motives and, anyhow, the best upon which we can rely for the meeting of this island nation's need for transfer points between sea and inland transport so long as the interests of each port remain distinct and competitive.

So, many millions of pounds have been invested in port structures and machinery, and each port is governed by a body whose prime duty is at least to earn a return on the money of stock-holders invested in that particular port, or to make the demand for rate-aid to meet a deficit as small as possible. Pursuing their separate aims, they may be led to engage in expensive competition to gain traffic from a neighbouring port, involving perhaps the provision of costly new facilities which, when completed, will render idle the like type of facility already installed at the port which loses the traffic. In other cases the free play of competition may force a particular port management to forbear to provide equipment which traders urgently need, because the cost cannot readily or wisely be incurred.

National Supervision of Ports.

Whoever studies the question impartially can hardly fail to see that to-day and, still more so, in the future, ports are facilities peculiarly needing to be controlled and aided, if not entirely managed, by the nation through an appropriate Ministry. Even people not prone to think this way have seen glimmerings of reason for thirty years. The Royal Commission on the Port of London early in the century, when discussing the great developments made by Continental

186

ports, chiefly owing to liberal State outlay, observed :
(at p. 444) :

> The power of undertaking large present expenditure,
> and of working for a long time at a loss with a view to
> compensation in a distant future, is no doubt, in the
> keen world competition, an advantage possessed by
> undertakings which have the force of an Empire, State
> or great city behind them. If, in some countries,
> national and municipal resources are thus employed, it
> becomes most difficult for private enterprise elsewhere
> to hold its own against the intelligent, far-sighted and
> formidable rivalry thus created.

Isolated Action Useless.

The Chairman of the Dominions Royal Commission,
when addressing the London Chamber of Commerce
in December 1917 on the improvement of sea communications, pointed out that so long as harbour
development was left to isolated and independent
action by individual harbour authorities, further great
progress could not be expected, as the responsibility
for the development of harbours along great ocean
routes must be dealt with by superior State authority,
acting independently of the immediate interest of
individuals.

In July 1930 Sir George Buchanan, K.C.I.E.,
approaching the problem as a port expert and not as a
politician, told the Institute of Transport that whatever
form of port control was adopted, the advantages of
a close relationship between the ports and the transporting media behind the ports were so great that
co-ordination by some common controlling authority
was of first importance. Sir George, who visited
Australia at the invitation of its Prime Minister to
furnish the Australian Government with the data on

which to base a national scheme, found that Australia did not suffer from a shortage of ports but that in their administration there were many defects, in that Port government was always subject to political influence, ports were in no sense self-contained financially, and there were strange anomalies in the port assessment. Sir George, however, reached the somewhat illogical conclusion not to favour nationalization or State management and control since, though a thousand reasons could be given in its support, on paper, the fact remained that political intrigue and interference could not be prevented. He favoured autonomous control by port trusts, but if the State had to advance capital for a new port under development, the State should guide rather than control. The port should be allowed a free hand in all matters of detailed working, but should consult the State on policy and important matters of development, and the latter should exercise a beneficent supervision over port finance ; but in no case should port revenues be incorporated with those of the State nor port funds be diverted for general State purposes.

Present Competitive Situation in Britain.

While not forgetful of the immediate advantages sometimes secured by British traders through the existence of the current fierce competition, and also from the continuance of such rather Gilbertian competition as exists between the Port of London Authority and the private wharves on Thames-side, it is submitted that the time is quickly coming when a measure at least of genuine national planning must be applied to ports. Others support this view. Sir Eric Geddes tried, in his Bill to Establish a Ministry of Ways and

Communications which, much whittled down, eventually became the Ministry of Transport Act, 1919, to secure powers on the lines now suggested. Indeed, it is noteworthy that, whereas in 1919 the Port of London Authority and other public port trusts vigorously opposed the grant to the new Ministry of the powers then sought, the Port of London Authority's General Manager now publicly advocates something of the sort after having investigated in person the methods employed in administering the South African Ports and Harbours which, with the country's Railways, are under centralized Government ownership and control.

Geographical Grouping of Ports.

Mr. R. J. Hall, a prominent Liverpool business man, has often urged a scheme for the grouping of our ports, and by way of example pays special attention to the Merseyside docks. The powerful Transport and General Workers' Union strongly supports the grouping of our ports, and detailed and highly interesting evidence on these lines was given to the Royal Commission on Transport by Mr. Ernest Bevin, who desired separate groups for the Humber, the North-east Coast, the Firth of Forth, the Clyde, the Mersey, and the Bristol Channel.

Mr. Frank Brown (Assistant General Manager, Port of Bristol Authority) has forcefully urged in public the need for port reorganization, pointing out how this country suffers because it was a pioneer. Heavy burdens are being carried in respect of capital expenditure on docks which are now out of date, because of the increasing size of vessels, the shrinkage in overseas trade and the changes in character of traffic since the

War. Since 1890 the capital expenditure on dock undertakings has been more than doubled, and interest has now to be found on well over £200,000,000 beside heavy annual maintenance expenses. These many separate ownerships, and the many and varied artifices to attract traffic, have led to certain trades being largely dealt with at certain ports, and so to much cross-country haulage and haulage for needlessly long distances, and to the handling, for example, of goods for Bristol through London or Liverpool, and vice versa. The sectionally minded transport man may like these long hauls, but clearly most of them are uneconomic and so against the nation's interests.

Mr. Frank Brown considers the whole problem of docks and their relation to ships and railways and overseas trade generally of such paramount importance as to warrant the prompt attention of an expert Royal Commission. Some references were made to the subject in the Final Report of the Royal Commission on Transport (Cmd. 3751 of 1931) but they skated delicately over thin ice and, as one expert said, knew neither the thickness of the ice nor the depth beneath. They thought railway ownership of docks undesirable (para. 479) and deemed the public trust the ideal owning body, and still more so a trust which controlled not just a single port, but all the harbours in a particular district. They were, however, uncertain whether to recommend that ports which to-day are railway owned should be transferred from their present ownership.

Would Railway Control solve the Problem?

Some students wish to see all ports brought under railway control, and there is no denying the feasibility of this solution, but it has many grave disadvantages. Unless the railways were first unified, preferably as a public trust, it is hard to see what real assurance there would be that road, canal and coastwise transport and their users could secure from the ports as square a deal as railway users. Even with the unified railways as the sole port authority of the country, those familiar with the details of the situation are very doubtful whether, as a general policy, the necessary equality of opportunity would be assured to all forms of internal transport. The argument that a port was, in effect, a much glorified goods station and could, in railway hands, be worked as such ignores the complex warehousing and expert operations which trade needs demand shall be properly attended to at the main housing ports.

One has admitted that the present port system may permit one individual at times to gain an immediate trade advantage at the expense of others ; but surely this should not prevent the long view from being taken ? Modern large-scale industrial groups are surely more concerned with permanent and substantial benefits, than with snatching temporary advantages at the cost of heavier outlays later on, and impoverishing other groups of consumers. Hence the least we can do is to work for a legal enactment grouping the ports, much as railways were grouped ; and it may well be sounder to jump a stage and propose a national direction of ports under an expert Board.

Which Ports deserve most Encouragement ?

But if we go on to forecast what a controlling body would do, and how a national plan for ports might be constructed, grave contradictory tendencies at once are seen and some compromise *must* be reached. The all-important decisions as to which areas shall be favoured for port development and which discouraged turn mainly on a study of existing facilities, a decision as to which are working with greatest efficiency, and which are becoming obsolescent. The work of regional planning authorities must be noted, and of any industrial siting board or like body, for drastic changes in industrial planning are probable as the full implications of the power age are realized.

Traders would wish the retention of as many facilities as possible and would like to find good port facilities as near as might be to every place from or to which they want to move goods. Working against this is the ship-owner's wish to be able to pick up or discharge his cargo at as few points around our coast as possible. Each entry into a port involves fairly heavy disbursements, so there is clearly a point in value of traffic handled, below which calls at several ports around our coasts are definitely uneconomic to the shipowner even if the port authority, in earnest but perhaps misguided effort to secure boats, quotes a very low rate of tonnage dues for a part cargo entry or clearance.

Conflicting Interests of Traders and Shipowners.

This rationalization of ports, then, may not all work one way. Traders may come to see that pressure on shipowners to provide multiple sailings or bring their boats to the point nearest their premises may not always be sound policy. If heavy and uneconomic outlay is involved which is reflected in higher dock charges and shipping freights, these may well offset the savings made by paying only for the shorter land transit. In the final analysis, to cut out calls by ocean liners at various smaller ports and to make greater use of coastwise and canal facilities for bringing essential foodstuffs and raw materials to their actual consumers may prove the better solution. One throws this out, without dogmatizing.

Never has there been any genuine attempt by the three main parties to this problem—traders, shipowners and port officials—to get together and reach a concerted policy to deal with the situation on a basis wider than that of the interests of a particular port area. How could there be, for to-day there is nobody at all empowered or equipped to take the broad national view of our port facilities and their development? One reiterates this statement, not forgetting the useful work of the Dock and Harbour Authorities' Association.

There is much in the shipowners' contention that multiple calls by a vessel around the coast may inordinately increase running costs—which increase, through Conference commitments, they cannot always pass right on to the parties who demand the calls. Opposed to this, under to-day's conditions of senseless struggle, is the pressure from firms or industries that cargo liners put into the port involving the shortest land or canal haul, so that traders may keep their throughout costs down. Who can arbitrate between them and reach a sane, measured conclusion in the true interests not only of the two contestants but of all the parties whose transport costs and competitive efficiency are indirectly involved?

If a limited view only be taken, this point seems a strong one for railway ownership of ports: for it is much easier for throughout costs to F.O.B. (or from C.I.F. on imports) to be adjusted to competitive needs even when a long land haul is involved when rail and port are under one control than when each is seeking an independent livelihood. Still, this again is a selfish and not a nationally helpful line of thought, and surely cannot stand against the much greater national benefits possible from a thoroughly determined overhaul of the entire problem.

Road Haulier Handicapped at some Ports.

The last point suggests another. The custom of making arrangements to merge rail and port charges in a single figure was sound when planned, but to-day it should come under searching review from some authority superior to the two participating bodies. Much traffic enters and leaves some ports by road, and the propriety of the continuance of agreements which markedly favour the rails as inland carriers is being increasingly called into question. At any rate most industrial transport officers hold that port authorities who are parties to these F.A.S. (i.e. free alongside) and like arrangements should be required to enter into comparable arrangements, so far as may be, when shipping traffic is carried by road. On the contrary, some railway-owned ports have tried hard to exclude the long-distance haulier altogether—another argument for national control of ports.

Port Charge Unification.

How far the unification of port charges, at which some reformers aim, would be made less difficult under these proposed conditions is not clear, for that desirable topic bristles with material complications. Some twenty-five years ago the then newly formed Port of London Authority sent an expert charges officer to visit other important British ports and prepare comparative statements of the charges in force thereat for six main import traffics. A comparison based on schedules alone, without knowledge of the actual services comprised in each rate, and the differing customs of the trades at each port, was clearly quite valueless—and still is—and this was intended as one of several preliminary steps to a scientific revision and unification of the port's charges. The more thorough this officer's researches the more complex was his problem seen to be. After months of expert work and the preparation of many pages of "double-brief" size comparative statements bristling with signs and footnotes, the task was set aside temporarily while the practical needs of the moment were met by quite empirical revisions of many of the tariffs.

London's port charges since 1909 have suffered many flat percentage increases and decreases, and many tariffs have been revised in consultation with trade organizations to meet special needs, but the scientific investigation designed to lead to the ascertainment of a new and presumably logical basis for the charges has apparently never been resumed. These facts are recalled not to discourage reformers but to give them a due sense of the difficulties to be encountered in this field.

Even so, there is no theoretical reason why it should be harder to arrive at a standard basis for port charges than it was for railway rates. The real problem lies far deeper. Port charges depend not on a mere "labour-and-materials cost" which might be averaged for all ports but, to a marked degree, on the heavy but vastly different overhead costs involved in the development and maintenance of each separate port, and on the radically different "customs of the port" affecting the discharge and working of cargo : customs which are clung to with amazing tenacity by one or another inevitably self-centred interest.

Fundamental legal changes are needed to ensure the pooling of physical resources and the simultaneous pooling of revenues to meet capital and sinking fund charges. Only bracketed with such drastic financial steps can one require port authorities to unify their main tariffs, but as part of such changes it would be feasible.

Bitter Legacy of Oppression of Labour.

The equally vital matter of securing the elimination of such diversity of customs of the port and labour agreements as operate to prevent the best results being secured from modern mechanical equipment, and trying to arrange definite and parallel lines of operation, calls for a very high level of statesmanship. Negotiation of these matters with the trades union is handicapped by the past history of more than fifty years of bitterness, during which the sorely oppressed docker succeeded, solely through self-sacrifice and organization, in wringing from his employers somewhat fairer conditions of service and pay. Unless they are given a really enlightened lead, and learn that they may respect and trust port managements, it may be expected that the grandsons of the poor men and women who struggled and starved to secure the "docker's tanner" (the rise from 5*d.* to 6*d.* per hour granted in 1889 after a most bitterly fought strike) will look with suspicion on plans to reduce labour, to modernize dock equipment and to use it with the maximum of economic effect. Though in the long run the workers may get part of the advantage from more efficient use of capital, such advantage is usually too remote for the worker to appreciate. It is useless for the industrialist to grumble at the reactionary attitude of the docker whilst he himself supports an economic system in which the few thrive and the masses struggle under conditions of deliberately continued shortage, and waste of food. But, treated with honesty by a newer generation of port officers, and as part of a general change in outlook toward the use of machinery in this age of potential plenty, the docker may see reason also and the sweeping away of restrictive agreements should be a central pivotal action leading to much more economical working at the ports and a levelling downwards of the costs. Obviously the movement towards turning the docker from a casual to a permanent worker, or, at least, a man with some security of employment, will facilitate these changes.

Costly Equipment Inadequately Used.

Only then is there any real hope for the achievement of the big economies possible under another subheading—those to be gained from the considerable lengthening of the total working day for ship-to-shore work and perhaps also warehouse work—by the adoption of shift working by the outdoor labour and supervisory staffs at ports. Mr. T. Bernard Hare studied this aspect of the subject closely and computed that only 31 per cent. of the time during which dock accommodation for handling cargo was available was actually used for revenue-earning purposes. Obviously if by a proper understanding with labour this proportion could be doubled, the capital invested in the facilities might be better remunerated, labour could get a share of the benefit, and before long the dock charges might be materially lowered. A fully equipped berth in a modern dock may cost anything between £350,000 and £750,000 plus heavy depreciation, maintenance and overhead costs which vary but little with the amount of use obtained from the accommodation provided.

Mr. Hare's figures for five large docks, generally considered to be so fully occupied that extensions were needed, showed that on an average each berth was in use for ten days and out of use eight days : i.e. there was no ship at the berth, though in some cases the transit shed would be occupied by goods for part of the time. Bringing this comparison down to hours, the proportion borne by the total berth hours to the hours during which loading and discharging were taking place was as 100 is to 23. Closer study of these statistics proves that the proportion of the time lost due to bad weather, machinery out of order, awaiting trucks, lorries or barges, waiting delivery instructions, etc., was so small as to be negligible. The main cause of lost time to ship and dock alike is the habit of only working eight hours out of the twenty-four.

Now, though it is inherent in some industries that the capital involved can only be in use for a small proportion of its time, the transport industry is surely among the last that need be put in this category : especially so since at the ports goods are transferred from the largest and most costly carrying unit—the ocean liner—and it is important to maintain continuity. Amicable arrangements for shift working to get at least 16 hours' usage daily out of ship and dock would be a most valuable achievement. Since many ports cannot now accommodate much bigger ships without constructing new docks, and since the already high cost of dock construction will increase rapidly as greater depths are necessary, the case for national study of the whole subject and the approval of a definite policy of port development is clear and action is urgent. Whatever is done must ensure that all facilities are used to their fullest extent.

Clearly, the really worth-while economies can only be gained if this problem be tackled at its root, with coats off and, more important, minds open to new ideas, and efforts are made (1) to reach an entirely new understanding with the labour force at work in ports, and (2) to vest the ports in some form of common national ownership. All efforts that avoid these hard truths are but scratchings of the surface and can give but niggling economies.

However, those not ready for anything so bold can amuse themselves working for some sort of National Port Advisory Board—possibly the super port authority of Sir David Owen. Some such national board for ports, previously suggested in 1930 by Mr. Ernest Bevin before the Royal Commission on Transport, will to-day come as less of a shock to industrial opinion. Some port officials themselves are also thinking that way and many who in 1919 strongly opposed Sir Eric Geddes' plans for the establishment of a Ministry of Ways and Communications are now, in essence, advocating what he had in mind.

Functions of a National Port Control Board

Mr. Bevin's National Port Control Board would not have anything to do with the management of the individual ports, but rather would be a survey board or a board which would take a view of the industry as a whole in so far as new developments and capital

expenditure were concerned : it would have some
check on unwise capital outlay and would survey and
determine the normal requirements of the country.
For this purpose the Board should always have before
it the country's full needs and be in constant touch
with the movements of industry and population, the
development of new industries, the requirements of
coastal trade and other water communications, and so
forth.

Frankly, this over-riding advisory control body
alone does not greatly appeal to me, for surely each
time it suggested anything of material value it would
be faced with the most vehemently expressed opposition, backed by political wire-pulling, from the
area which was required to curtail an activity or
change a policy in the general interests : and naturally
so, since that authority's duty to its shareholders under
the present regime might be interfered with by the
action proposed by the Control Board.

Clearly, there is no real way out but to follow the
" working model " of a merger of competing transport
interests already provided by the London Passenger
Transport Board, and surely when conversions are
as rapid and thorough as those touched on above, one
need not in 1935 apologize for outlining still more
thorough-going proposals. Yet to console the apostles
of gradualness, the partial step is mentioned, and they
can get pleasant revolutionary thrills from helping
to bring it about. Unless vaster changes are actually
in progress, their first objective should be the appointment of a Royal Commission to investigate and report,
without delay, whether grouping of ports by estuaries,
or the formation of one public port trust would be
the more desirable from the viewpoint of public

interest. The main purposes to be attained are well set out thus by Mr. Frank Brown :

(1) The discouragement of such competition as takes merely the form of diversion of trade without public advantage.

(2) The co-ordination of port activities so as to minimize conflicting services and ensure that the whole machine may work more smoothly, economically and efficiently.

(3) The simplification of charges, both in respect of dues on ships and goods, as well as for the processes of loading and unloading, the straightening out of the tangle between railway rates and services, and dock rates and services.

(4) The better provision for the amortization of capital represented by obsolete or obsolescent dock accommodation and facilities.

(5) The encouragement of traffic to flow—as far as other economic factors permit—along natural channels.

(6) The elimination, in the interests of trader and dock worker as well as the public, of customs and practices which are restrictive and inimical to the efficient and economical conduct of business.

CHAPTER TWELVE

COASTWISE SHIPPING'S NATURAL ADVANTAGES

THE British coastline is pierced at irregular intervals by the estuaries of rivers and on them—Southampton Water, Thames, Humber, Tees, Tyne, Forth, Tay, Clyde, Mersey and Severn—lie our greatest ocean ports. On the east coast between the ocean ports are situated many lesser ports so that Great Britain, for her size, possesses more ports than any other industrial country. Indeed, so situated are they in relation to the industrial areas that it has been calculated that 80 per cent of the country's inhabitants live within 15 miles of the sea.

Yet many people know little of the British coasting trade, in which are engaged over 1,200 vessels with a gross tonnage of over 800,000, maintaining 308 services around the coasts (excluding many services from Glasgow, Oban and Fort William to the Western Islands of Scotland). London is served from all the major ports and many of the lesser by direct boats. Through Manchester, Liverpool and Hull the towns of industrial Yorkshire and Lancashire have their egress to the sea and cheap transport to every other area of the country. The Bristol Channel ports serve the thickly populated area of South Wales, and Glasgow, Leith, Dundee and Aberdeen cover the hinterland of Scotland.

Beside the liner services, coasting tramps are always available to the shippers of bulk cargoes for the feeding of one district with the produce of another. Of the total coastwise tonnage, 60 per cent is coal. Thames-side cement is carried all over the kingdom by coasting tramps. They bring stones and slag from the north and the west to the road-making material plants. They carry slate from North Wales, tin plates from Swansea, china clay from Cornwall to ports which serve the Potteries, transfer boilers and machinery from one shipyard to another, distribute salt, pig iron, steel rails, steel billets, grain, potatoes, sugar, burnt ore and timber. With no regular route, their services available to anyone with bulk cargo to move, they knock about the coast from Lerwick to Penzance, from Grimsby to Tralee from one year's end to another.

The coasting trade offers direct employment to some 12,000 seamen and a much greater amount of indirect employment. The coaster is of great value in redistributing from, and feeding to, the ocean ports. Where large quantities are concerned, the coaster can moor directly alongside the ocean vessel and save dues and charges by making a direct overside transfer. With smaller parcels a transfer by barge can often be arranged at lower cost than if a landing and re-loading operation were involved.

Coaster's General Utility.

The coaster may not be suited to certain perishable or urgent goods, but for a large proportion of our manufactures, foodstuffs, raw materials and bulky cargo of low value from port to port and hinterland, the coaster offers a complementary and alternative form of transport to road and rail : and even, in time

203

of emergency, an essential service. Spokesmen for the coasting trade do also make a strong claim that, like the canals, they are, for traders, an important safeguard against monopoly—to some minds an ever-present danger. Our coasting trade has suffered severely from foreign competition. It is open to the ships of all nations. There is nothing to prevent a foreign vessel taking a cargo of salt from Weston Point to Wick, a cargo of burnt ore from London to the Tyne, or a cargo of china clay from Cornwall to Kirkcaldy. Pressure for the reservation of our coasting trade did not succeed, owing to fears of retaliation which would be unfavourable to the shipping industry as a whole. So motor-driven coasters, operated mainly by Dutch owners, are familiar sights in our ports. Useful craft of light draught and good speed, not needing firemen and, indeed, often " manned " by the master's wife, family and relations, these vessels render good service but undoubtedly have harmed our British coasting trade. British owners are now building ships of similar type, but since they must be operated in compliance with Board of Trade regulations, their costs are likely to stay higher than the Dutch operator's costs.

Railway and Coastal Rivalry.

Competition between the coasting trade and the railways has at times been severe, but they have now arrived at certain " conference " arrangements on lines somewhat like those agreed between the canal interests and the railways. Yet competition has not ceased, for each month the railways make or revise some exceptional rate or other and give the Railway Rates Tribunal as their reason " water competition ".

Coasting trade experts who have complained of

railway rate-cutting point out that in some cases railway companies carry at below cost and, to make up their loss of revenue, make higher charges between points where the coasting vessel does not offer an alternative service. From places adjacent to ports, say within a radius of 20 to 30 miles, it is said to have been the habit of the railways to cut the throughout rate, while maintaining the standard rate on the short haul to the port in an endeavour to take traffic away from the coasting ships. Such efforts have not always succeeded : they have reduced the coasting steamer's rate ; at times they have forced the coasting steamer to bring the traffic to the port by road ; and the net result is that the railway companies carry their share of the traffic at a lesser rate than they need do. The coasting concerns do not feel that the Railways Act, 1921, actually gives them the protection it ostensibly did against such competition, nor do they feel very satisfied with the safeguards introduced into the Road and Rail Traffic Act, 1933, at their request.

Coasting Trade Wants Better Berthage.

The coasting trade have a strong case for greater consideration in the matter of port facilities. Charges at railway ports still average 60 per cent above pre-War and seem to be maintained at this level in the interests of railway policy rather than in the national interests. In ports under public control, fairer conditions prevail, but in many ports, both publicly and privately owned, there has been and still is a tendency to relegate the coasting trade to the more antiquated parts of the docks. Though in the past anything has been good enough for the coasting trade, conditions are certainly improving.

This matter of berthing is, of course, of primary importance to the coaster, and the ideal is for the vessel to be able to arrive at and depart from her berth at any state of the tide. When a service is operated to a fixed time table, as the coasting liner services are, it is not only irritating, but wasteful, to have a ship lying idle for several hours owing to the state of the tide. There are ports where the tides do not cause delay, as either deep water berths or lockway systems are available. When improvements are being planned, port authorities might more earnestly consider how to help the coasting trade by providing more " open quay " facilities.

Then, too, the dock labour difficulties sometimes are especially sharply felt by coasting trade owners and consignors. As an expert writer put it, " The atmosphere of suspicion and class war is very prevalent in some areas, and the grievous sins of previous generations of employers are being visited upon successors who often have a more enlightened outlook." In some ports work cannot be continued after certain hours. If by fog or storms a vessel has been delayed, it may not be possible to get her back into turn by crowding the work of two days into a day and a night, even if the owner is quite willing to pay the higher rates naturally sought for such work. Thus the coaster's reputation for reliability tends to suffer and some traders do not place in the services the confidence that, on the whole, they probably deserve.

Unhappy Legacy of Suspicion.

The same unhappy past accounts for the undoubted suspicion of to-day's labour, and reluctance to countenance the introduction of labour-displacing machinery—for, under the present system, that is what it is. Without a new system which will make it truly "labour-saving" machinery not much hope is here to be seen.

The student of general economic problems may like to know that whereas in 1728, over 77 per cent of vessels arriving in London were coasters; by 1860, coastwise and foreign trade vessels practically equalled each other in numbers, and when 1933 figures are examined it is seen that the balance has swung round again so that one gets the following curious division on percentage lines :

					Vessels. %	Tons. %
With cargoes :						
Foreign	.	.	.	.	. $33\frac{1}{2}$	70
Coastwise	.	.	.	.	. $66\frac{1}{4}$	30
With cargoes and in ballast :						
Foreign	.	.	.	.	. 26	60
Coastwise	.	.	.	.	. 74	40

Heavy Coastwise Ton-Mileage of Freight.

A director of a leading coastwise shipping company recently calculated that the 30,000,000 tons of cargo carried coastwise during 1933 travelled an average of at least 250 miles, so that the coastwise ton-mileage that year would be at least 7,500 millions as compared with the total railway freight traffic ton-miles for the same year of 14,000 millions. It may surprise many to learn that the coastwise freight ton-mileage, as thus estimated, is over half that of the railways, and the calculation does emphasise the national value of this service which at so small a capital outlay does an important job of work.

Use Nature's Free Track.

Another important comparison has been produced for the confounding of critics of coastal shipping. It is pointed out that of the total railway capital over £900,000,000 had been spent in constructing the permanent way and the annual outlay to-day upon the maintenance of the track, plus signalmen's wages and equipment, is £21,000,000. These figures represent over £1,000 per mile of track or 1s. 8d. per ton on every ton of traffic carried (or, after making all allowance for passengers, 1s. 1d. per ton). The Salter Conference assessed the annual cost of British roads at £60,000,000. Yet the *total* capital invested in coastwise shipping is not over £20,000,000 and its track is for the most part laid free by nature.

Further practical difficulties faced by the coasting trade are the phenomenal increase in lighterage costs (London : 1913, 2s. per ton, and 1934, 3s. 6d. per ton ; while Hamburg in 1913 was 1s. 1d. and in 1934 was 1s. 4d. to 1s. 7d. per ton) and a still greater proportionate increase in handling charges at British ports (averaging in 1913 2s. 0½d. per ton and in 1934 5s. 7½d. per ton or an addition of 175½ per cent to 1913 figures).

The post-War coasting vessel is speedier than that current in 1913, but more costly. Running costs, plus depreciation, came out nearly double the pre-War level although the quantity of bunker coal was only increased by a few tons. In 1913 the order of expenses was : (1) handling cargo, (2) crews' wages, (3) bunkers, (4) depreciation, (5) port charges, (6) maintenance and repairs, (7) insurance, (8) management, (9) running stores. By 1934 handling charges (stevedoring) had greatly increased, as had port charges which, item (5) in 1913, had now gone up to second place in the list of expenses. The percentages compare as follows :

	1913	1934
Handling Charges :		
Half cargo on board . . .	31·20	46·83 + 15·63
Full „ „ „ . . .	41·70	59·49 + 17·79
Port Charges :		
Half cargo on board . . .	9·18	14·50 + 5·32
Full „ „ „ . . .	7·77	10·52 + 2·75

New Spirit of Enterprise.

The diesel engine was in 1934 responsible in part for a marked advance in British coastwise shipping. Shallow draught craft so equipped have remade ports such as Norwich and Totnes which cannot be reached by sea-going vessels drawing much water. A few years back Norwich had almost ceased to be a port, but in 1934 about 600 vessels passed up the Wensum carrying general goods and from thirty to forty thousand tons of coal. During the past few years sea-borne deliveries of coal to the river Thames have increased by one-half, and rail-borne tonnage has declined by one-third. About 12,000,000 tons of coal are carried annually to the Thames by sea, while rail carryings are 2,000,000 tons less than they were five years ago. Coastal freights are so low that coal is being brought from the Tyne to London for 2s. 6d. a ton or thereabouts ; the collieries of Northumberland, Durham, Fife and the Lothians are near tidewater, and on freights of this order they have a distinct pull over the Midland and Yorkshire pits in regard to the " delivered " price of coal in the London area.

Scope for More Train Ferries.

A word about Train Ferries, which so appeal to some lay minds that they would like to see many more. The advantages in time and cost of running goods through from country to country without breaking bulk are material, and handling costs at the ports are saved, but Customs requirements are not relaxed because the goods are in a wagon instead of a ship so that goods often have to be disturbed and re-loaded. Thus the purely artificial barrier detracts once again from the big advantages offered us by the engineer. Unless ideal but somewhat improbably bold measures of stock standardization on both sides of the Channel and North Sea are introduced, however, there is always the problem of special train ferry rolling stock. If existing and projected ferries are to fulfil their function the wagons must be readily available in any part of the country, or else the advantages of the through run will be lost and wagon-loads will have to be transhipped at the ports. The Customs must be used, too, to facilitate instead of retarding these modern services. Subject to these practical difficulties, however, there are great possibilities about the increased use and larger number of train ferry services between Britain and the Continent.

CHAPTER THIRTEEN

BETTER UNDERSTANDING WITH OCEAN SHIPPING INTERESTS

MUCH is amiss with our British ocean shipping industry ; so much that a more innocent writer might find
cause to wonder at the assurance of spokesmen for
the industry who move fervid toasts in its honour.

Not that shipowners as a class are worse, indi-
vidually and morally, than many other classes of
business men ; but they do succeed in shedding
tears so easily, to order, that they gain quite an
amount of undeserved sympathy from the public.

The poetic pæans to the bravery of our mercantile
marine, the warmly expressed admiration for our early
merchant adventurers and chartered companies—who
probably get, in retrospect, more bouquets than a
shrewd contemporary study of their motives and
political methods would show to be deserved—both
are apt to become confused in the minds of the public
with genuine altruistic enterprise. People tend to
forget, and by poetic counter-propaganda are helped
to forget, the slave traders who waxed fat, the coffin
ships and the long struggle for the establishment of a
Plimsoll Line, the general record of bad treatment
of men while profits were being raked in. Even in
1935 some very serious comments were heard in the
High Courts as to the foundering in mid-Atlantic of

certain vessels which apparently complied with the literal requirements of the Merchant Shipping Acts.

Shipping does seem a clear case where the excellent work of the inventive engineer is not yet being put to proper use by the administrator; or where the latter, sometimes struggling in the grip of an unhappy heritage, sees the true steps necessary to put the industry to rights, but through weakness or respectability fears to come out frankly on the side of thoroughgoing reform.

In boom years of the past profits have been divided by some concerns up to the hilt. Unwise or unconcerned managers ignored the fact that each year their fleets were growing older and in time would need to be replaced. Shareholders took all that was offered them by way of dividend, but when their capital had petered out, and liquidation of the concern was inevitable, they were apt to be very angry.

Great Advance in Technical Efficiency.

The story of the technical improvement of seagoing vessels would occupy many pages. Efforts have been made successfully in all directions to obtain lowered working costs. Even in the past few years there has been a 30 per cent all-round reduction in the cost of transporting a given amount of dead weight cargo at a given speed. A striking saving of 25 per cent on fuel alone has been made. A typical 8,000-ton steam cargo vessel, which just after the War could do 9 knots and consumed 25 tons of coal daily, can now do the same daily performance on about 18 tons of coal. Other fuel systems have also their advantages, though expert opinion differs on their comparative merits.

Yet British managements are full of complaints, and State help of one sort or another is sought; which seems rather like pleading for the nation to come to the help of one more class of private enterprise when it has made a mess of things for itself. Sir Archibald Hurd (a famous student of shipping matters) pointed out in May 1934:

> A country which subsidizes its shipping is rather like an amateur who, without knowing the leads of the wiring, endeavours to connect himself up with a telephone exchange. As the result, he gets a bad service himself and spoils the service provided for other people. Amateurs in the shipping industry—for Governments are always amateurs in any industry, however far into their taxpayers' pockets they may dip—can never provide the services which experts can provide except at a higher cost.

Shipowners' Muddled Logic.

Curious, if that is so, to reflect that a Government could easily hire shipping and other experts and could even remunerate them, if it chose, according to the success or otherwise of their department! Earlier in the same address Sir Archibald Hurd reminded hearers that during the Great War the movements of British ships were controlled by the Ministry of Shipping and their earnings were restricted to what were known as " blue-book rates " of hire—" rates many times less than those obtainable in the open market ". He does not complain that the " blue-book rates " were unremunerative to owners, so that the " open market " appears to have fleeced those who had to resort to it. Apparently, then, the taxpayer need not necessarily be stung?

Further along Sir Archibald treated us to a delicious morsel :

> Moreover, after the War, British owners, at the direct request of the Government, relieved the taxpayers of Government-built and ex-enemy tonnage at prices which to-day seem fantastic. It was the price they paid for freedom at a moment when leading politicians were talking of the nationalization of shipping.

And now, one supposes, having taken their profits on the use of those vessels with both hands during the boom years—profits which might otherwise have gone to the nation—there is a desire among ship-owners for help from any convenient quarter. It was revealed at the first meeting on 15th May 1935 of the contributories of the White Star Line that the company's total deficiency, including share capital, was £11,280,864. Gross liabilities were estimated at £2,643,670 and assets to realize £362,865. The senior official receiver said the company was promoted by Lord Kylsant in conjunction with the Royal Mail Steam Packet Co., and was registered in January 1927 with a nominal capital of £9,000,000. The statement of affairs showed that £2,621,398 was due to unsecured creditors, and attributed the company's failure to excessive prices agreed upon in respect of acquiring certain assets ; and heavy interest charges incurred in the acquisition of those assets, as well as the extra-ordinary slump in shipping. This was a creditors' liquidation, and there was no chance of the share-holders receiving anything back.

Though perhaps the most striking recent example of a terrible state of affairs it is by no means the only one. Some hope may be seen in the gradual rise to power of a newer and more enlightened generation of shipping trade leaders, but even they are very badly overdue and very much in danger of succumbing to *force majeure* within their organizations and families, and settling into a defensive battle against the Government and their clients instead of bestirring themselves boldly to make the revolutionary changes in plan and thought which are the chief need of the industry. The one really bright spot is the ready effort to promote and meet public demand for more varied holidays afloat—though it took a landsman to initiate that idea.

International Difficulties Facing Shipowners.

The main trouble lies in the grave difficulties of securing international agreement on shipping matters ; whereon such agreement clearly is much more vital than in internal matters. This difficulty forms an inevitable and potent alibi which is persistently used every time bold plans are pictured. This prevents such a control of shipping routine and charges as is now recognized as nationally necessary for rail and, in a measure, road transport, and permits the shipping fraternity to maintain their attitude of reluctance to negotiate with their clients or to heed their advice.

Shipping freights are not published for the general information. They are fixed in secret by officials of the Shipping Conferences, and circulated for the confidential information of the constituent members of those particular Conferences. Seldom, if ever, do the tariffs get into the possession even of shipping agents unless they are authorized district representatives of a particular line—and that system is passing. Much less likely is it that a copy of such a tariff will fall into the hands of a merchant or manufacturer—except by unofficial and perhaps unworthy surreptitious means. There is no place to which a trader can go with a hope of finding anything like a complete set of shipping tariffs exhibited, nor has any publisher yet had the hardihood to try to compile a guide or reference book of any kind of shipping freight classifications and tariffs. Why is this, and is it sensible?

It springs from two main causes. First, simply because shipping is international. *Each* nation knows that, and its handling of the shipping interests is probably the more cautious and ineffective because of the knowledge. Although on matters of design, treatment of men, and safety, public opinion is active and has pressed certain reforms upon the shipowners, the questions of classification and freights hardly excite the same clamour, and so there is not enough impetus to force any Government to intervene seriously in these directions. Hence freight rate problems, and the adjustment of interests between shipowners and clients, have never yet seriously engaged the attention of civilized Governments as has been so with railways and their rates. This, maybe, has another basic reason. Prominent and pioneer shipping companies appear to have been courted by Governments for their value in developing foreign trade and bringing traffic which could pay customs duties. Mail subsidies and other favours were showered upon them, and the whole psychology was widely different from that which has long subsisted between railways and Governments.

Merchant Shippers losing their Key Position.

Secondly, foreign business has largely developed in the hands of the merchants—merchant shippers and merchant importers—rather than in those of the actual users of the imported products and makers of the goods for export. Though the merchant houses did pioneer work which has played its part in Britain's commercial development, they are not primarily concerned about shipping freights nor even in the discovery and pushing of British goods to the exclusion of anything else. Now that exporting is passing more under the direct control of British manufacturers and they are getting a grip of the freight situation, they find themselves in serious difficulties when they enter on negotiations with the shipping companies. There is even in some cases a reluctance to admit their status in the matter and their right to negotiate. Of course, those concerned are doing their best to alter this stupid state of affairs, and shipping companies are gradually giving way. The pity is, though, that there should be any need to have to force them to a process of " giving way ".

This is not an attack on the general level of freight rates—which, speaking generally, is a low one—but it is a frank criticism of what has been referred to publicly as the " stone-walling, secretive and injudicious manner of handling shipping matters which appears to be common to all the steamship lines ".

Exporters Advised to Organize.

As far back as December 1922 the Imperial Shipping Committee strongly recommended that for every trade route the shippers should form representative associations to meet the Conferences, their already constituted counterparts among the shipowners, to discuss and settle outstanding differences and questions of mutual concern as they arose " and in particular to the exercise of the option, if desired by any appreciable proportion of the trade, between the rebate and agreement systems and the negotiation of the specific terms of such systems ". This "deferred rebate" system by which a supplementary percentage on the freight is collected by the shipping company from each shipper and retained from 6 to 12 months, then being returned on application so long as the shipper has continued to send his goods only by the Conference lines, has always come in for much criticism, and alternative plans have been pressed for.

Except in the North Atlantic trade there has been practically no response to this suggestion. Even in that one trade the Canadian section of the North Atlantic Westbound Freight Association (the shipowners' Conference, that is) recently issued a form of agreement for signature by shippers under a penalty of an increase in freights of 50 per cent, without consulting the body of shippers they "officially recognized ". Both the Australian Traders Association and the South African Traders Association are dominated by the merchant shipper. It is demanded that each Conference should be willing to meet a representative body of manufacturer shippers, but abundant evidence exists to show that hitherto the Lines have shown extreme reluctance to entertain the idea.

Tariff Secrecy should be Abolished.

Freight tariffs should be published, and all who will pay a reasonable sum to cover publication costs should be entitled to receive a copy. Some traders are certainly paying needlessly high freights solely because they are not describing their goods in line with the terminology adopted by the Conferences. Those familiar with the General Railway Classification will appreciate the point.

Deferred shipping rebates have been made illegal in several countries, and back in July 1922 the Chamber of Shipping of the United Kingdom told the Imperial Shipping Committee that shipowners were prepared, in their several conferences, if satisfied that the substantial majority of shippers in number and volume so desired, to discuss with shippers the precise terms upon which the option of the alternative net freight system could be given. Yet shipowners have shown little disposition to meet the reasonable requests of manufacturer shippers in this matter. The suspicion is that some merchants and a certain type of forwarding agent derive substantial benefits from the deferred rebate system; yet why should the Lines acquiesce in such a position? Even shipowners themselves are not free from criticism here. Many small manufacturers do not understand that rebates are due to them, or omit to claim them, so that if they do not pass into the pockets of merchants or forwarding agents they are retained by the shipowners. The Lines have no moral claim to this money, and one is tempted to inquire whether there is a vested interest in the retention of the deferred rebate system.

Shipowners' Mentality Unchanged!

A number of grievances which affect certain groups of manufacturers could be outlined. The situation respecting Heavy Weight surcharges led a prominent shipper to state that the " mentality of the shipowners had not stirred from the days of *Cutty Sark* " since they ignore all the advances in ship's gear and port cranes and insist on imposing a " heavy weight surcharge " on pieces exceeding 2 tons in weight. The Railway Companies' heavy weight surcharge scale starts at 12 tons; surely shipowners should raise theirs to at least 6 tons? Still more iniquitous is the practice of certain Conferences which raise their heavy lift charges on a measurement instead of a weight basis.

Reform in Passenger Fares.

A case which will be even more striking to the general reader can be made when we turn to passenger traffic problems and especially those of the North Atlantic trade. Major Frank Bustard, O.B.E., for many years a leading official on the passenger side of the White Star Line, has put his finger on a vital point to the future well-being of the world, when he points to the highly unsatisfactory situation for passengers crossing between Europe and America. Major Bustard doubts the likelihood of the resumption of a heavy emigrant traffic to America, and urges that the future of the North Atlantic passenger trade is bound up with business and tourist travel—but of these, he says, only the surface has yet been scratched. Such travel ought vastly to increase—language, friendship, business interest, all demand it—but the great ban is that the cost to-day is out of all proportion to both the pre-War fares and the fares charged for sea transport on cruises and to other parts of the world.

The travelling public now regards £2 per day as a fair rate for cruise travel, and many highly attractive cruises are offered at even less than £1 per day. Yet North Atlantic passenger fares on certain first-class ships work out at £8 to £10 per day, and even third-class rates during the summer are from £2 to £3 per day for very inferior accommodation. Despite the many reasons advanced for the retention of this high scale the public remains convinced that the fares are excessive, out of date, and restrictive. The *real* reason the lines decline to lower rates is their difficulties because of the changed conditions and the fact that they continue to use out-of-date methods and employ unsuitable vessels. When there was a big flow of emigrants they paid the running costs of the " monster " boats and the ultra luxurious facilities for first-class travellers could be " pyramided " upon the lower classes. With emigrant traffic virtually dead the future of the North Atlantic trade lies with vessels of a more moderate tonnage taking not more than 1,500 passengers, such as the *Georgic, Britannic, Manhattan, Lafayette,* all built in recent years, all comfortable—indeed, luxurious—and all showing satisfactory voyage returns even to-day. If the expensive, extravagant and out-of-date " monsters " could be laid off, passenger rates could be materially reduced and the number of travellers greatly increased.

If the slavish retention of an extravagant scale of victualling that makes of every meal a banquet could cease, and more up-to-date methods, both in feeding and decoration, were instituted, there could be materially lowered costs.

Business-like Project Discouraged.

Yet we find that a project to secure and operate ships on the North Atlantic at reasonable prices was frowned on by the Treasury and the Bank of England, most unfairly and short-sightedly, because financial aid for the completion of the *Queen Mary* was being given. What a stupid situation for the exponents of private enterprise. Get in a sufficiently bad way and the State will then back you and will see that nobody competes with you ! But, with great self-denial, it will maintain the sanctity of private enterprise in that it will fail to seize the obvious opportunity to take you over at market price and operate your plant successfully in the national interests. What a muddled state of mind arises in the Government from subservience to the money grip and a pitiful readiness to obey its irresponsible and selfish dictates.

The simple truth is that shipowning concerns for the most part need the sharpest jerks to bring them up to date with the public's demands and ready to meet them in a commercial spirit. Inherently sea transport is cheap, and if it cannot be operated with economy the main faults lie with the direction. The Annual Reports of the Chamber of Shipping make most pitiful and plaintive reading. The speeches at Shipping Companies' Annual Meetings are equally pitiful when one pierces through the brave show of words in search—unsuccessful search—of signs of a master mind which has grasped the realities of its chosen job in life. The only negative sign of sense gleaned from them is the reluctant admission that laws to confine Empire trade to Empire vessels, or reserving our coasting trade, would not, on balance, help them, for from foreign retaliation they might lose more than they would gain. Nor is much hope to be seen in the present Government.

Why not National Shipping ?

But, one submits, there is no reason why an enlightened Government of the people for the people should not operate ocean liners with success and true economy, and this would be the way to real reform. A purchase by the exchange of stock at market prices, and rationalizing of the principal liner services and their operation on commercial lines in the nation's interests, would now not be difficult. Then there would be room for considerations of true human value, and, by the big increase of travel on such journeys as between America and Britain, the friendship of the nations would be more firmly forged and the peace of the world would be helped.

To suggest that State ownership has not been successful is to misread history—a mistake for which the average man can hardly be blamed since the true facts are hard to come by. I dare assert, however, that study would reveal that those instances usually quoted as examples of the failure of State ownership are not that at all, but are rather proofs that gigantic robbery of the nation can go on almost openly when a Government put in power by the money interests is eager to do the bidding of its true masters. The trend of this argument can only be illustrated by the observation that an undertaking to purchase vessels at " production cost plus a percentage " may encourage those sharing the " percentage " to take a more lenient view of the cost sheets than they otherwise might do.

CHAPTER FOURTEEN

TRANSPORT'S TRUE PLACE IN AN AGE OF PLENTY

THE reader must have realized that amidst the diverse selection of facts and opinions here brought together —and the selection might have been much extended— certain broad tendencies and certain plain economic facts stick out a mile. But from those facts and tendencies very different sets of inferences might be drawn.

Let me artlessly assume that the function of any Government, and therefore of our Government, is to look after the welfare of the people of its country ; since in theory the Government consists of elected representatives having delegated duties. Let me further suppose our Government seriously intends to tackle its tasks. He who sees irony here merely dis- closes the tortuousness of his own mind or the grave disillusion of his spirit.

The hoardings testify to the present Government's satisfaction at having so well accomplished its tasks, and its hopes for another lease of power to accomplish some more. But among dubious voices we hear that old fighter, Mr. Lloyd George, at Cambridge (August 2nd 1935) thus : " The whole of their attitude towards unemployment is one of self-satisfied and arrogant impotence. Let us have an end of all this gaping and yawning and strutting in front of a great task. The nation must insist on its being tackled." Many others are equally dissatisfied.

Any Government that seriously attends to its business must put the true interests of the entire population before the selfish interests of limited groups. It must do some very remarkable things. Chiefly, it must plan ahead and work consistently to its plan, instead of making hasty and illogical decisions on isolated points, when pressed hard by one interest or another. For " planning " is now in the air. We cannot escape it if we would. People who trembled at the idea because it suggested " Russia " are now uncurling again and beginning to murmur the word for themselves.

Transport's Integral Place in National Planning.

If we are to make the most of this age of potential plenty, and fully and equitably to use the marvellous powers and gifts of the machines, the much more skilful planning of the growth and use of each part of our national resources becomes imperative. And the solution of this chief problem will carry with it the solution of all our most urgent problems—health, housing, unemployment, economic rivalry, war and all the related ills and difficulties.

A Government sincerely trying to do its duty to the majority of its constituents would give early and thorough attention to Britain's current transport problems and future needs. True, our Governments for many years past have partially recognized the pivotal national value of transport facilities, though efforts to trace any consecutive and constructive policy in these matters only succeed when historians are of the conventional " whitewashing " type. Even then one suspects them of much after-wit if not of a thorough-going sublimation of motives.

Certainly Governments, our own not excluded, have given some serious thought to the military aspects of transport, and in that regard have at times exercised foresight. How tragic to reflect that military needs have so often hastened inventive progress ! Research is readily financed during war-time, but at other times Governments are niggardly except over preparing for some new war.

Not that official support for research work is often needed in transport. Confidence that good pioneering work would be recognized and applied, rather than stultified, would alone do wonders in that direction. Government's real function should be the active direction of policy in the operation, firstly, of all existing transport services of a public or semi-public nature, and then their complete re-moulding more adequately to meet our greater needs.

" Planning " does not mean " Cutting Down ".

For " planning " does not mean " cutting down ". It means applying all our resources to subserve the purposes of a fuller life for everybody. It means organizing the production and distribution of a much bigger volume of foods and manufactures ; radically improving the standard of housing comfort ; steadily reducing the amount of work regularly required of each able citizen ; and hastening the development and wider use of the newer amenities of life.

This is the complete and only intelligent answer to the worker who to-day is tempted to sabotage progress, and to the skilled technician and organizer who clearly

226

sees that the better he handles his job the faster is he putting himself and others out of a job. Unless these vital elements of our population can see the real goal, and can trust the honesty of their leaders, how much longer can they be persuaded to be fairly willing accomplices at their own destruction?

This would appear to mean the imperative demand for an entire change of Government policy, and the reversal of all current efforts designed to find ways of limiting, impeding and stopping transport in the interests of one private group or another. So long, of course, as personal activities must continue to be restricted when they conflict with the community's welfare.

We are an Open-air People.

In seeking the basis for our " new age " let us realize that though economic factors change materially, psychological factors change but little. Preference for open-air pursuits and rural surroundings has persistently dominated our race and has brought us back, though towns have been forced on us by mis-understood economic pressure, to the ideal of the garden city and suburb. It seems we have never really wanted the city, and that may be why our cities are, from the town planning viewpoint, so bad. Other nations have delighted in the city, and the pride of the citizen is reflected in its structure and embellishments. We remain, at backbone, an agricultural and open-air loving people, and the remedy now rightly becoming popular is to provide increasingly for the population on lines that at least simulate the countryside—shorn, some say, of its inconveniences. This means rapid transit facilities and the increasing dependence on

227

road transport to take people and goods where railways cannot well penetrate.

Agriculture can hardly be displaced from its basic position, but it can now be improved in technique, and those who make it their individual life's work can at least be saved the worst rigours of physical hardship and isolation.

Financial Reform the Key to the Position.

An entire change in our system is needed, so that what is physically possible shall be made financially possible. I belong to the quickly growing school of those who have seen the flaws in the system of scarcity economics in which they were trained, and who refuse to permit starvation amidst plenty. Whether the detailed technique proposed for this purpose by Major C. H. Douglas is the best or not, is not here material. What matters immensely is that the Douglas Social Credit analysis, like other plans of similar import, pierce to the root problem of to-day, for they reduce money to its real place as a token which may conveniently represent true wealth, but is not itself true wealth. By their simple common sense these proposals to expand and regulate the supply of money tokens, so that they equal the supply of real wealth, must come quickly to the front and, as the multitude grasp them, such a howl of rage will go up that radical changes in our financial system will be inevitable. Money and goods will be equated, and things to-day said to be " economically impossible " since they do not show our financial dictators enough easy profit, will be fully possible because our true wealth will allow them to be provided and, even as now, fresh facilities will beget fresh demands.

228

But even before these changes dawn fully on us, and while we are still burdened with leaders whose minds are atrophied by fear, and whose consciences are dulled by habitual respect for " number one ", much can be done to bring our transport facilities into line with current needs. Nobody can be excused from thinking out a National Transport policy because he is just waiting for bigger changes. The way to put things right is to proceed at once on the next step.

Instruments for National Transport Supervision.

Indeed, dawning recognition of the need for a national transport policy of a bolder type is seen in the recent creation of a Transport Advisory Council, a body appointed by the Transport Minister after public agitation and a Royal Commission report, and charged with duties which (given enough courage and vision) they may interpret as of a fundamental nature. There is also now a separate advisory committee on road safety. These bodies are doing something, but there is need of publicity for their work.

What emerges from this study is something much more drastic and, some may feel, more terrible. It is that true co-ordination and ordered development of our transport services can best be secured by the early creation of a National Transport Board superior to each and every service directorate, to take control of the whole problem. The members of this Board, who should be appointed on the nomination of the Minister of Transport, should be full-time officers and have no extraneous interests. Subject to Parliamentary approval the Minister should lay down the general policy of the Board, but the Board should be responsible for applying this policy. The Board must not be large, but must be large enough to maintain liaison and effective contact with sub-sections each responsible for a particular arm of transport ; and the members must be appointed for their technical knowledge and business ability. Adequate attention to the users' as well as the operators' side must be a feature of the appointments.

Unification by way of exchange of stock holdings should proceed by such stages as the Board itself might determine. The speed of the changes would depend upon the political colour of the Government, but the general trend is inevitable. Only such a Board, taking over and extending the functions of the Transport Ministry to-day, would be in a position to bring about the really worth-while economies and improvements which the public desire, and reconcile the apparently conflicting interests simultaneously.

Basic Data for Transport Board.

Before the activities of such a National Transport Board could be detailed, its future work would have to be connected clearly with the community's needs, and with the tasks being grappled with by other planning Boards. Such a Board, to be able to plan effectively, would have to heed some important factors to which to-day the quarrelling transport units do not pay any consistent regard. For example, the Board should know for what future population to lay plans. Changes in population due to births and deaths can be fairly accurately forecast ; those due to migration cannot—unless you control the migration. To-day this is not done, for there is no official policy as to how we want our people distributed, nor any machinery for such control. To supply these missing cogs is a big task, yet without them planning, in a real sense, is impossible.

Responsible town planners advocate controlling town development by grouping our population into towns of moderate size ; not running into each other, but so arranged that all can have easy access to ample playing fields and a real stretch of open country. Resistance comes only from those who refuse to give due weight to human values. As Mr. F. J. Osborn, of Welwyn Garden City, puts it :

> If the stupidest or most corrupt committee were got together and empowered to decide how towns should develop, they would never have the hardihood or cruelty to build London or New York as they are to-day. These cities are not triumphs of civilization ; their form is no part of the permanent achievements of science and culture. . . . Such things spring out of a long history of superstition and distorted psychology.
>
> Human beings can accept them. But no committee of human beings, starting from scratch, would dare perpetrate them, so condemning millions to spend ten hours a week like rats in underground pipes ; hundreds of thousands of children and old people to be cemented up on the fifth or tenth floor of tenements, with no outlook but chimney pots and other tenements. It is only because nobody does it, because London and New York just happen, because there is nobody to be shot at for the fantastic folly of it, that humanity has accepted without revolt hitherto these dreadful gifts from the Gods or Devils.

Large-scale national planning will be bound to stop the present form of growth of large cities. It will have to reach conclusions as to the sort of town development that best suits modern industrial, commercial, cultural and social needs, and in exercising control the planners must determine to foster towns of the right size and structure and to decentralize those of excessive size and concentration. Thus they would quickly have to reach logical conclusions as to the optimum sizes and densities of towns to fulfil the various urban functions, and the general sort of position such towns should occupy in relation to the rest of the country.

Control of New Industrial Location.

Without a national plan the general drift of haphazard development will always be toward centralization, making town conditions grow steadily worse. But there is nothing inevitable in that drift. The settlement of new industry and new commercial premises is the largest factor in the growth of towns. The inception of new industrial enterprises in London makes possible, if it does not actually cause, the immigration of people from other parts of Britain. Lack of new businesses leads to the decline of certain old towns and the creation of " derelict " areas. National control over the settlement of new industries will check the growth of a town already large enough. It is the master key to national territorial planning, as Mr. Osborn points out. So a vital part of the work of the national planning body will be that part carried out by a National Siting Board in concert with that of a National Housing Board and, when in full swing, this work will for the first time permit prediction of the growth of population and of new activities in each regional and town planning area. Especially it will enable over-centralized towns to get to work on their built-up areas, and maybe abolish the terrible dilemma which now confronts London—the choice between tubes and tenements.

232

When there is a National Transport Board working in liaison with the other planning authorities, it will for the first time be possible to take a truly national view of the community's transport needs. Then only will waste elimination not be held up by vested interests which think they thrive from competition—a view now passing from the minds of transport leaders. The nation then could plan and develop its transport services boldly. The fear that initiative would be stultified is ill-founded, for each nationally owned branch of transport could be inspired by technical and administrative officers of high quality and even greater initiative than is possible in small competitive units struggling for quick profits. There would become possible, especially in the managerial grades, periodical interchange of staff to promote wider knowledge and help get rid of the sectional mind.

Only then is there any prospect of true co-ordination, implying the use of each type of transport service always and for every purpose for which it is the most truly efficient, while gradually eliminating wasteful overlapping—all this without prejudice to the true interests of the shareholders in existing undertakings, even the " widows and orphans " so often and plaintively trotted out.

How will Transport Board Function?

Just what decisions a National Transport Board would take in particular circumstances cannot yet be forecast, though sketchy outlines have been ventured upon earlier, so far as they spring logically from the basic considerations set forth. The main thing to remember, however—this to reassure the timorous— is that the changes here advocated will come as part

233

of still bigger changes in our way of living, leading to more leisure for the individual and a much greater volume of traffic to be moved. Thus, national ownership and control of transport will not mean a limiting or cramping control, but rather a skilled planning of the direction and character of fresh development, or substitution of greater for less efficiency in a given arm of the service.

The views of national ownership and control of transport held by the minority political parties have never been disguised, but it is significant that at least fifteen years back the first Minister of Transport was responsible for plans which visualized that the State, on behalf of the people as a whole, had a moral claim to share in surplus railway revenues in exchange for a material enlargement of their original charters. This precedent, as well as the limited terms of the original railway charters, must be kept well before the public even though it be not now thought worth while to challenge the propriety of those original charters.

Public Money for Public Benefit.

On new and even more cogent grounds the right of certain transport services to keep all their earnings may now be seen to have been forfeited. Why should the nation make grants and guarantees of interest to private concerns? All the big new works which have been achieved or begun by various transport authorities with the incentive of Government support or guarantees, which under the ostensible aim of providing work have materially added to the value of their properties, could just as well have been constructed without incurring a penny of debt to the financiers. Only a sane system of money tickets, as

234

tried with success in Guernsey a century ago, was needed. Thus public support really amounts to a documentary expedient, planned by and for the money power, to seize the chance afforded by the difficult times into which that power itself had plunged us, to put that power still more firmly in command by assuring it a further volume of perpetual toll from the operators and users of the transport service so " aided ".

Thus, in the financing of the recent large-scale canal improvements what really happened is that the Government promised, if necessary, to hand over £500,000 of your and my money to the financial groups who, by creating credit out of book entries, have kindly allowed that valuable new work to be undertaken. They have thus nearly doubled, to you and me, what was the true cost of doing that work. In essence, the financing of the £40,000,000 London Passenger Transport Board scheme is the same, thus showing that such a Board, though a big step forward from chaos, has yet to overcome its chief obstacle before it can give the public a fair deal.

Why perpetuate Grave Errors?

Leaving aside finance for the moment : in our transport set-up we have suffered through being pioneers. Yet need we go on handicapping ourselves for the grave errors of our forefathers ? British railway history might have been quite other than it has proved to be, but though we inherit there a burdensome legacy from the past, we need not impose new burdens on the railway or any other systems of transport. One cannot respect a mind that says, " We are handicapped in such and such ways ; therefore handicap our competitors too." *All* transport systems should be

235

relieved of needless burdens and allowed to operate to the greatest of their powers in the public interest : but that view will not get enough general support until all are publicly controlled. Then only can we get heed to the basic view that transport as such should not be taxed at all. It is basically wrong to penalize a public need in that way. And clearly the imposts falling on road transport have long passed the stage when they are needed to meet the costs incurred through such transport.

The viciousness and inequality of present taxation on road transport is apparent when it is realized that fuel oil used by the coastwise trade and by vehicles of the " railcar " class escape the tax which falls on road vehicles so operated. No taxation on transport should do more than barely cover the administrative and track constructional costs clearly incurred by that class of transport.

Honest Treatment of Workers.

Though the Labour Party is still tragically short-sighted in that it has not yet got beyond the stage of struggling for a better distribution of a limited supply of money tickets, instead of working for a rapid expansion of that supply so that no class shall benefit at the expense of another ; yet that party and its adherents must be seriously reckoned with in any scheme of transport reform. " Ca canny " and even deliberate sabotage cannot be far away unless the justifiable suspicions of the workers can be frankly faced and removed. This means, for the moment, that the results of more efficient methods and machinery *must* be more fully shared with the partially displaced labour, and as its right, not as a grudging charity. The conception of "legacy from the past" cunningly introduced by the railways into the minds of the Salter Conference, must be expanded so that we see that current scientific achievements are not the private property of the few but, being the results of the work of many generations of pioneers, now dead, are the true and inalienable legacy from the past from which all have equal claim to benefit.

Points that some readers may have looked for in this book have barely been referred to. Much has been written since the Weir report on general electrification of main-line railways, and the reader may have expected a definite expression of opinion on that topic. It has not been given, partly because it would involve the statement of too many preliminary technicalities to set out and defend a reasoned conclusion : and partly since the true point at issue to-day on that matter is not technical, but financial. Under Social Credit, electrification would be easy and useful ; without it, the chances of making large-scale railway electrification pay are doubtful, and probably the piecemeal achievement of the plan, in sympathy with the movement of population, is saner. Similar reasons account for other notable omissions from this book.

Groping amongst Inadequate Data.

On other points one has had to write cautiously and avoid too definite a statement. Such is the controversy as to whether the railways do actually carry their heavy mineral traffics at a loss. No member of the public has access to all the facts—nor have many railwaymen—and it may be they are not accurately known by anybody. Debaters are therefore forced to rely on biassed statements and incomplete figures.

Only when by a common national ownership or at least a co-ordinated control of all chief means of transport the reason for presenting tendentious figures has gone shall we secure data, not now existent, on which the truth of such statements can be known and so the sane line of policy can be determined.

Similarly, on that vexed matter of railway capital it is possible that, among the thousands of words one has read and heard, the full truth, shorn of interested claims, has not been grasped. Maybe in that instance the whole truth will never be known, for the problem will presently be empirically settled by a series of bargains between the contestants. If that be likely, I would suggest that railway shareholders impel their directors to enter upon the bargaining process quickly, for they may never again be in such a strong bargaining situation as they are to-day—despite all contrary semblances.

Of course, the readjustment of railway capital would not of itself enable the railways to earn a penny more cash ; *but* it would remove their main propaganda argument of poor dividends, show their true working results and put things in better perspective, so that even if carried through under the present regime it would re-orientate all the railway officers' minds.

Re-establish Public Confidence.

Another important step which does not depend on a change of ownership of transport services but could be carried through at once, is the issue of a public statement showing the conditions of the establishment of the Road Fund, and of a plan for restitution to the Fund of the sums improperly seized by the Exchequer. This robbery of the Fund is a factor sapping the public trust in the fairness of political administration. Restitution would help restore that faith that a bargain made would be kept, and would enable most of the big and needed road improvements to be started without delay, and so make for an entire change of mental outlook among all branches of road users.

A point, perhaps theoretical, which also must be set right is the impression sought to be fostered that road transport should contribute to the amortization of the initial costs of the roads. Not only is this unfair when regard be had for the important community use of roads and the unearned increment their existence has brought to many frontagers : but it is definitely opposed to the growing appreciation that both aircraft services and ground organization for aircraft quite properly deserve some Government aid to establish them properly.

The Central Waterway Board, the National Port Board, the National Airways Boards, and Boards governing the Coastwise Shipping, Railways and Long Distance Road Transport (both passenger and goods) will each form integral sub-committees of the National Transport Board. Ocean Shipping, now organized in two national bodies, would do well to merge these bodies and put them in shape also to act as a sub-committee of the National Board. Otherwise, without too much regard to the overseas selfish interests of some of our shipowners, a more rigid control may be imposed in return for the assistance now sought.

With such an organization, co-ordination between the types of service would be possible and the shrinkage in volume of staff through administrative economies should for several years be met by a steady increase in volume of traffic movement to a point more nearly approaching the true needs of the people. Long before that point was finally reached, we should have learned the true value of leisure, and should have realized that leisure does not necessarily make a man a libertine or a drunkard. Our fears for the moral character of the other man will have proved needless.

Toward a " World Commonweal ".

While this progress is being achieved, we must forge and intensify our contacts with technicians of other nations, carrying through as many as possible of our progressive measures in full accord with transport experts of other nations. Good results are already attending efforts of this sort and, as H. G. Wells shows in " What are we to do with our Lives ", while no static Utopia is now conceivable, we are forging the machinery of the future World State. Though Wells deems its development inevitable and imperative, he points out that possibly the whole idea of a " State " will undergo material changes in the human mind. The World State will not be a single central controlling body, making arbitrary decrees and enforcing them on reluctant constituents by armed force, but will consist of a whole series of agreements and conventions, reached voluntarily and for good business reasons by industrial, commercial and technical workers in each of the diverse spheres of activity that make possible our civilization. That is why the term " World Commonweal " is better than " World State ", for it more fully and happily connotes what men of wider friendship are earnestly seeking.

For Product Safety Concerns and Information please contact our EU
representative GPSR@taylorandfrancis.com
Taylor & Francis Verlag GmbH, Kaufingerstraße 24, 80331 München, Germany

www.ingramcontent.com/pod-product-compliance
Lightning Source LLC
Chambersburg PA
CBHW050420280326
41932CB00013BA/1929